Oxford International Primary Science

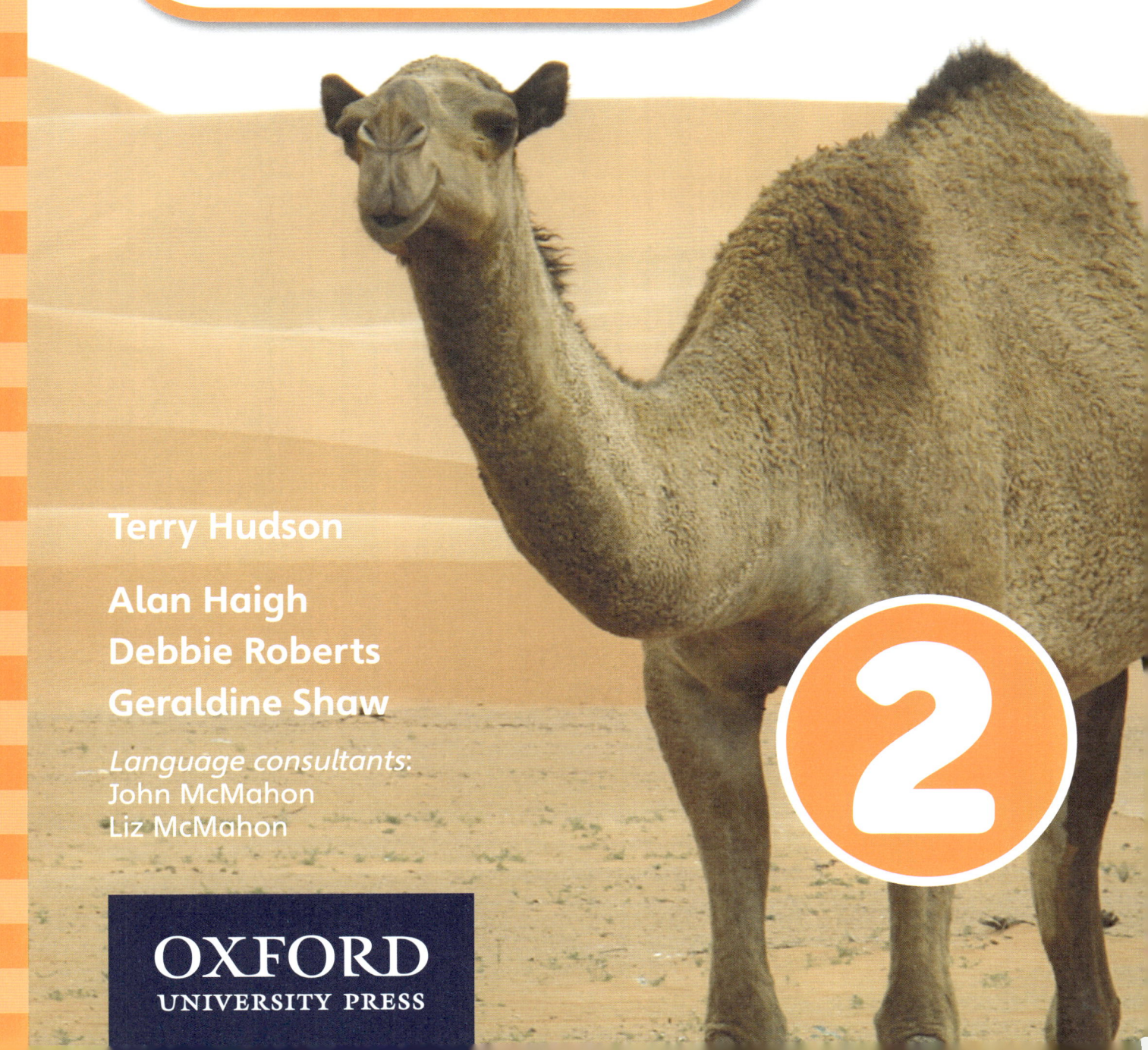

Terry Hudson

Alan Haigh
Debbie Roberts
Geraldine Shaw

Language consultants:
John McMahon
Liz McMahon

2

OXFORD
UNIVERSITY PRESS

OXFORD
UNIVERSITY PRESS

Great Clarendon Street, Oxford, OX2 6DP, United Kingdom

Oxford University Press is a department of the University of Oxford. It furthers the University's objective of excellence in research, scholarship, and education by publishing worldwide. Oxford is a registered trade mark of Oxford University Press in the UK and in certain other countries

© Terry Hudson, Alan Haigh, Debbie Roberts, Geraldine Shaw 2014

The moral rights of the authors have been asserted

First published in 2014

All rights reserved. No part of this publication may be reproduced, stored in a retrieval system, or transmitted, in any form or by any means, without the prior permission in writing of Oxford University Press, or as expressly permitted by law, by licence or under terms agreed with the appropriate reprographics rights organization. Enquiries concerning reproduction outside the
scope of the above should be sent to the Rights Department, Oxford University Press, at the
address above.

You must not circulate this work in any other form and you must impose this same condition on any acquirer

British Library Cataloguing in Publication Data
Data available

978-0-19-839478-5

10 9 8 7 6 5 4

Paper used in the production of this book is a natural, recyclable product made from wood grown in sustainable forests. The manufacturing process conforms to the environmental regulations of the country of origin.

Printed in Great Britain by Bell and Bain Ltd, Glasgow

The questions, example answers, marks awarded and comments that appear in this book were written by the author(s). In examination, the way marks would be awarded to answers like these may be different.

Acknowledgements

The publishers would like to thank the following for permissions to use their photographs:

Cover photo: Saudi Desert Photos by TARIQ-M, P4_5: Stringer/Getty Images News/Getty Images ,P5a: Mike Liu/Shutterstock, P5b: Dante Fenolio/Photo Researchers/Getty Images , P5c: Ian Schofield/Shutterstock, P5d: TK, P6: Shutterstock, P7a: Shutterstock, P7b: Yuriy Mazur/Fotolia, P7c: Fotolia, P7d: Shutterstock, P8a: Kongsak/Shutterstock, P8b: imagebroker / Alamy, P8c: Gordon Bell/Shutterstock, P10: Robertus Pudyanto/Flickr Vision/Getty Images, P13: Robbie Shone/Aurora/Getty Images, P14: Shutterstock, P18_19: Shutterstock, P18a: Eric Audras/Onoky/Corbis, P18b: Stuart Monk/Shutterstock, P20a: Shutterstock, P20b: Shutterstock, P20c: Shutterstock, P20d: Shutterstock, P21: Business Wire, P22a: DK Limited/CORBIS, P22b: Africa Studio/Shutterstock, P22c: Visuals Unlimited/Corbis/Image Library, P26: Digital Storm/Shutterstock, P28: Gaby Kooijman/Shutterstock, P34: Andy Crawford/Getty Images, P38_39: Thirteen/Shutterstock, P39a: REX/Steve Davey, P39b: Vyaseleva Elena/Shutterstock, P39c: Shutterstock, P39d: discpicture/Shutterstoc, P40: Vigya Pant/Bhaskar Chandra, P42a: photosync/Shutterstock, P42b: pling/shutterstock, P43a: Shutterstock, P43b: Qpic-Images/Shutterstock, P43c: Alex Kuzovlev.Shutterstock, P44: KIM NGUYEN/Shutterstock, P46: Yehuda Boltshauser/Shutterstock, P48a: silver-john/Shutterstock, P48b: Alexandra Lande/Shutterstock, P48c: ANDREW LAMBERT PHOTOGRAPHY/SCIENCE PHOTO LIBRARY, P48d: Imageman/Shutterstock, P48e: MARTYN F. CHILLMAID/SCIENCE PHOTO LIBRARY, P49: Iakov Kalinin/Shutterstock, P50a: FhF Greenmedia/GAP Photos, P50b: Shutterstock, P50c: Shutterstock, P50d: Shutterstock, P50e: Mayovskyy Andrew/Shutterstock, P50: Sergey Goruppa/OUP, P50g: Hellen Sergeyeva/Shutterstock, P50h: Madlen/Shutterstock, P50i: Maglara/Shutterstock, P54_55:Bbbar/Dreamstime.com, P55a: Eric and David Hosking/CORBIS,
P55b: Shutterstock, P55c: Shutterstock, P55d: Donna/Bigstock, P55e: Dieter Heinemann/Westend61/Corbis, P55f: Alexander Kataytsev/Dreamstime.com, P55g: Zhiqian Li/Dreamstime.com, P56a: Shutterstock, P56b: Zdenek Chaloupka/Shutterstock, P56c: Hung Chung Chih/Shutterstock, P57a: Shutterstock, P57b: Momatiuk - Eastcott/Corbis, P57c: Julian Calverley/Corbis, P58a: Vitalii Nesterchuk/Shutterstock, P58b: kavram/Shutterstock, P58c: Seqoya/Shutterstock, P60a: Siim Sepp/Shutterstock, P60b: Siim Sepp/Shutterstock, P61a: Jane Rix/Shutterstock, P61b: Pablo Hidalgo/Shutterstock, P61c: Glenn Young/Shutterstock, P62: Dave G. Houser/Corbis/Image Library, P63a: Auris/Dreamstime.com, P63b: Shutterstock, P63c: Tezzstock/Dreamstime.com, P63d: Borja Andreu/Shutterstock, P63e: marekuliasz/Shutterstock, P64: B. Anthony Stewart/National Geographic/Getty Images, P66a: Zoran Simin/Shutterstock, P66b: Oliver Sved/Shutterstock, P70_71: ScreenDress/iStock.com, P71: pixelparticle/Shutterstock, P72: Alastair Wallace/Shutterstock, P76: Lane V. Erickson/Shutterstock, P82_83: PLANETARY VISIONS LTD/SCIENCE PHOTO LIBRARY, P83a: JI de Wet/Shutterstock, P83b: Photodisc/OUP ,P84a: Corbis/Digital Stock/OUP, P84b: Vadim Petrakov/Shutterstock, P84c: leospek/Shutterstock, P84d: Louis-Marie Preau/Hemis/Corbis/Image Library, P84e: Sergio Schnitzler/Dreamstime.com, P84f: Michael & Patricia Fogden/CORBIS, P84g: SARIN KUNTHONG/Shutterstock, P84h: Leena Robinson/Shutterstock, P84i: Joseph Scott Photography/Shutterstock, P84j: Fedor Selivanov/Shutterstock, P84k: Deyan Georgiev/Shutterstock, P84l: Vladimir Voronin/Dreamstime.com, P84m: Tatiana Belova/Shutterstock, P84n: Christian Musat/Shutterstock, P85a: Elgru/Dreamstime.com, P85b: M.Khebra/Shutterstock, P86a: Chaikovskiy Igor/Shutterstock, P86b: Patrick Poendl/Shutterstock, P86c: BOB GIBBONS/SCIENCE PHOTO LIBRARY, P86d: DAN SAMS/SCIENCE PHOTO LIBRARY, P87: Lenar Musin/Shutterstock, P88a: oriontrail/Shutterstock, P88b: Vladimir Wrangel/Shutterstock, P88c: Michael Kern/Visuals Unlimited/Corbis/Image Library, P89: Chekaramit/Shutterstock, P90: Rostislav Glinsky/Shutterstock, P91a: Richard Whitcombe/Shutterstock, P91b: Paul Aniszewski/Shutterstock, P91c: Zack Frank/Shutterstock, P91d: Leon P/Shutterstock, P91e: Daniel Handl/Shutterstock, P92a: Ayman Aljammaz/Flickr/Getty Images, P92b: Fayez Nureldine/AFP/Getty Images, P93: Cubo Images/SuperStock , P94a: NASA image courtesy of Jeff Schmaltz, MODIS Rapid Response Team, NASA-Goddard Space Flight Center, P94b: Yuriy Kulik/Shutterstock, P95: Shutterstock, P96: Issam Rimawi/ZUMA Press/Corbis, P97: mario.bono/Shutterstock

Although we have made every effort to trace and contact all copyright holders before publication this has not been possible in all cases. If notified, the publisher will rectify any errors or omissions at the earliest opportunity.

Links to third party websites are provided by Oxford in good faith and for information only. Oxford disclaims any responsibility for the materials contained in any third party website referenced in this work.

Contents

How to be a Scientist 2

1 Light and Dark 4
Where does light come from? 6
We need light to see things 10
What is a shadow? 14
What we have learned about light and dark 16

2 Electricity 18
Simple circuits 20
Designing, constructing and predicting 30
What we have learned about electricity 36

3 Changing Materials 38
Changing the shapes of materials 40
Heating materials 42
Cooling materials 46
Dissolving materials in water 48
Natural or not natural? 50
What we have learned about changing materials 52

4 Looking at Rocks 54
What are rocks? 56
Different building materials 58
Types of rock 60
How do we use rocks? 62
Different types of soil 64
What we have learned about looking at rocks 68

5 Day and Night 70
The Sun appears to move in the sky 72
Shadows change during the day 76
The Earth is spinning 78
What we have learned about day and night 80

6 Plants and Animals 82
Plants and animals in hot countries 84
Plants, animals and their environment 90
Weather and water in hot countries 94
What we have learned about plants and animals 100

Glossary 102

How to be a Scientist

Scientists wonder how things work. They try to find out about the world around them. They do this by using scientific enquiry.

The diagram shows the important ideas about scientific enquiry.

- **Start here** — Asking questions
- I think that …
- I am going to …
- I am looking for …
- I have found that …
- This means that …

2

An example investigation: Investigating dissolving

Asking questions

How can you ask questions?

Start your questions with words like 'which', 'what', 'do' and 'does'.

- Which solids dissolve in water?
- What happens to the solids when they are added to water?

I think that …

A prediction is when you say what you think will happen.

Question

Which materials dissolve in water?

Prediction

Salt and sugar.

I am going to …

When you plan an investigation think about how you will make it fair.

What will you keep the same?

- The amount of water.
- The amount of solid.

What will you change?

- The material you are testing.

I am looking for …

You will look closely to see which mixtures still have solid at the bottom and which ones do not.

I have found that …

There are many ways to record results. A good way is to complete a table. A table keeps results neat and tidy. It can help you to see patterns.

This means that …

Look at your results carefully. Compare the materials. Which seem to have disappeared? Which are still easy to see?
Was your prediction correct?

1 Light and Dark

In this module you will:
- find out about different light sources
- understand that darkness is when there is no light
- find out about shadows.

💬 The sailors are looking for people in the water. Why are they using a searchlight?

Word Cloud

light
look dark
guess (predict)
torch Sun
prove

Physics

Many animals have eyes. These let the animals see things.

🗨 Some fish that live deep in the ocean do not have eyes. Why?

Amazing fact

Light travels at 300 000 000 kilometres every second. That is very fast!

Light and Dark

5

Where does light come from?

Discover different light sources including the Sun.

Lightning is a source of light

The Big Idea

Some objects give out light and some do not.

A source of **light** is something that makes its own light.

This man is trying to read his newspaper using lightning as a light source.

💬 Why is this not a good idea?

✏️ Look at the words in the word bank. Fill in the boxes to show three things that are better light sources for the man.

Word Bank

Torch   Candle   Shiny spoon   Flower   Mirror   Light bulb

✏️ Write down the name of an important source of light that is not in the word bank.

Some sources of light are natural like the **Sun**, stars and volcanoes.

| Sun | Stars | Volcano |

Some things made by people give out light. These sources of light are artificial.

We use artificial sources of light at night or in **dark** places like caves. Electric lights and **torches** are artificial light sources.

Some natural sources of light, like the Sun, are brighter than others.

⚠️ It is dangerous to look at the Sun. It is so bright that it could blind you.

✏️ Why do people wear sunglasses?

Where does light come from? (continued)

It is night-time in this picture, but the people can see each other.

💬 How can they see each other? Where is the light coming from?

Parts of the town square below are in the light. Other parts are in the shade.

💬 Where is the light coming from? What is making the shade?

There are some shiny things that we think are sources of light, but they are not. They are bouncing the light from another source. This is called reflection. Look at the buildings below. The shiny surfaces are reflecting the light.

💬 Does a mirror give out light?

✏️ We can see our reflection in a mirror. Does this mean the mirror is a source of light?

✋ Investigation: Is a reflection a source of light?

1 With a partner, take it in turns to hold a mirror in front of your face to look at your reflection.

2 Your teacher will change the amount of light in the room.

Predict what you will see.

💬 What happens to your reflection?

There are many things that appear bright, but they are just reflections of light from somewhere else. They bounce light from a source into our eyes. The Moon and shiny objects behave like this.

✏️ Look at the pictures. Read the words and circle the sources of light.

Underline the ones which reflect light.

| Planet | Sun | Mirror | Spoon | Candle |

✏️ Name two natural sources of light.

✏️ Name two artificial sources of light.

Now turn to page 16 to review what you know.

We need light to see things

Know that darkness is when there is no light.

The Big Idea

Nobody can see in the dark.

When the Sun goes down it becomes dark

At night there is not a lot of natural light. It is dark.

✏️ What can you see in a dark room?

✏️ Write down three artificial sources of light that might help you to see well.

When the Sun goes down, or you switch off a light, there is no longer a source of light. We call this darkness. The light can no longer send information to our eyes so we find it difficult to see. This can cause us lots of problems. How could you test this?

✋ **Investigation: Do we need light to move around safely?**

1. Tie a blindfold over your eyes and try to move around the room.
2. Ask your partner to help you to stop bumping into things.

💬 Describe how this felt.

⚠️ Be careful when you are blindfolded. Walk slowly with your arms in front of you.

Light and Dark

11

We need light to see things (continued)

In darkness we rely on other senses like hearing. This helps us to move around safely and get to the places we cannot see. How could you test this?

✋ Investigation: Does seeing help us to hear better?

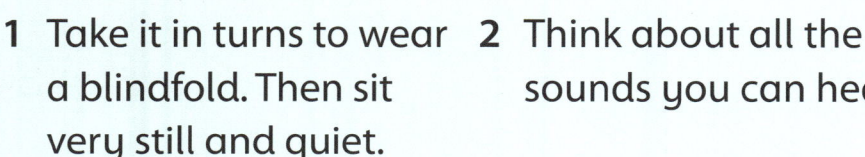

1 Take it in turns to wear a blindfold. Then sit very still and quiet.

2 Think about all the sounds you can hear.

3 Take off the blindfold. What can you hear now?

💬 When could you hear the most noises? Why?

Some people say they can see in the dark. This is not true. There has to be some light to see by.

Our brain makes our eyes get used to less light and it helps us to see things when there is very little light. How could you prove this?

✋ Investigation: Is all light the same?

1 Try to read this page in the dark.
2 Then use candlelight.
3 Now try it with a torch.

⚠ Don't touch the candle once it has been lit. You could get burned.

💬 What did you find? When was it easier to read the page?

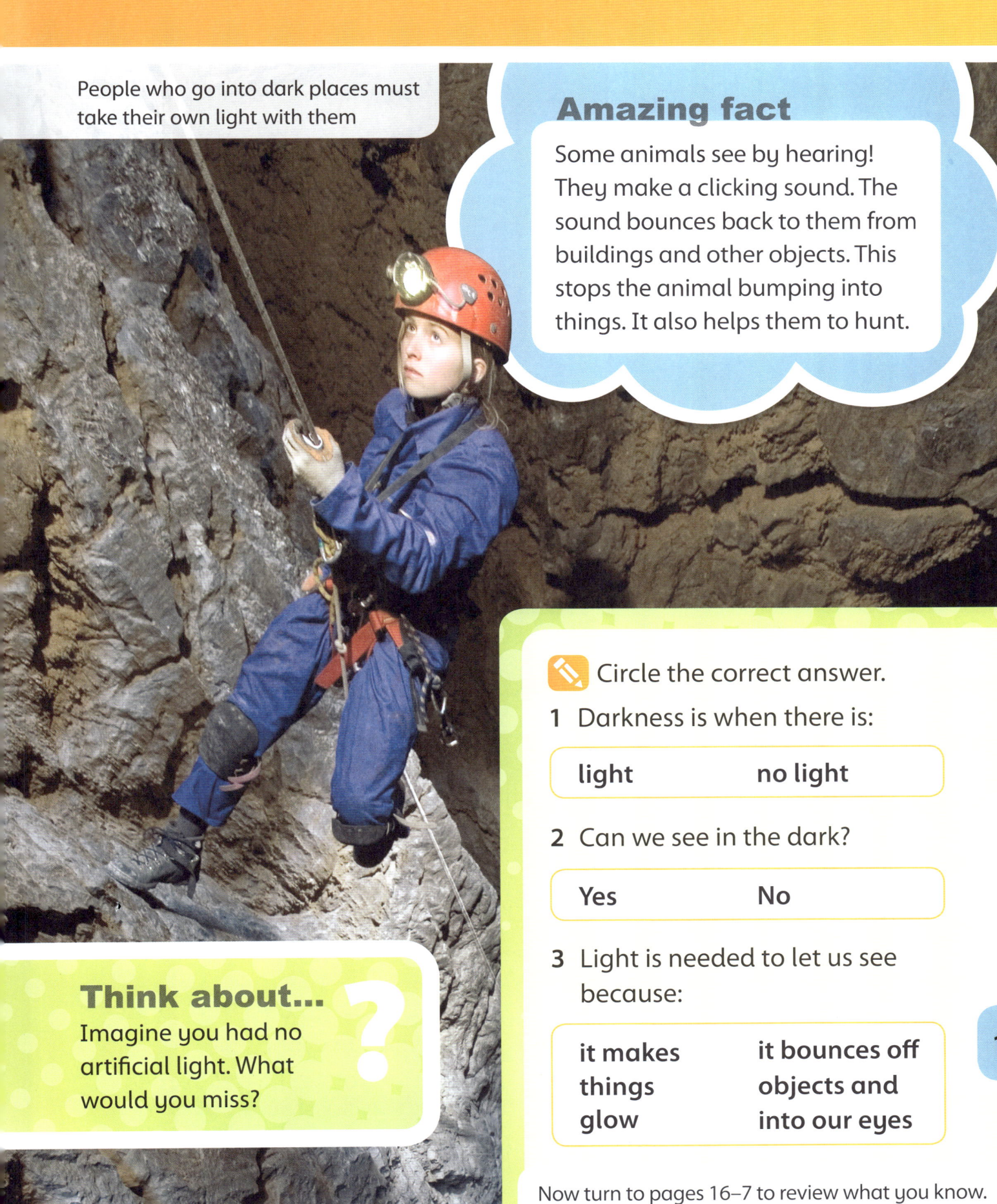

People who go into dark places must take their own light with them

Amazing fact

Some animals see by hearing! They make a clicking sound. The sound bounces back to them from buildings and other objects. This stops the animal bumping into things. It also helps them to hunt.

Think about...

Imagine you had no artificial light. What would you miss?

✏️ Circle the correct answer.

1 Darkness is when there is:

| light | no light |

2 Can we see in the dark?

| Yes | No |

3 Light is needed to let us see because:

| it makes things glow | it bounces off objects and into our eyes |

Now turn to pages 16–7 to review what you know.

Light and Dark

13

What is a shadow?

Understand shadows.

The Big Idea

Shadows are made by blocking light.

✏️ What can you see when it is dark?

✏️ During the day, what is the natural source of light?

When light from any source is blocked by an object it forms a dark patch. This is called a shadow.

When any source of light like a light bulb is blocked it forms a shadow. It is not just light from the Sun that makes a shadow.

💬 Try to find as many shadows as you can in the classroom.

💬 Have you ever noticed your own shadow?

✋ **Investigation: Using your hand to make a shadow**

1. Stand in front of a wall or clean surface.
2. Hold your hand in front of the torch.
3. Try making different shapes with your fingers.

✏️ If you hold the torch very close to your hand what happens to the shadow?

✏️ What happens if you hold the torch further away?

Think about...
A shadow is the same shape as the object that makes it.

✏️ Circle the correct answer.

1. Shadows are only made by the Sun.   True   False
2. Shadows are formed when an object blocks light.   True   False
3. Shadows never change size.   True   False

Now turn to page 17 to review what you know.

What we have learned about light and dark

Where does light come from? (pages 6–9)

✏️ There are two kinds of sources of light. One is natural like the Sun. What is the other kind of light source?

_____

✏️ When light bounces off a shiny surface like a mirror it is called …

_____

I understand the difference between natural and artificial light. ○

I know that shiny surfaces are not sources of light. ○

We need light to see things (pages 10–13)

✏️ Deep underground it is completely dark. Why is that?

_____

✏️ When we see things it is because _____ from the things has reached our eyes.

I understand that when there is no light it is called darkness. ◯

I know why we cannot see in the dark. ◯

What is a shadow? (pages 14–15)

✏️ Circle the correct answer.

1 Shadows are only made by the Sun.	True	False
2 Shadows are formed when an object blocks light.	True	False
3 Shadows never change size.	True	False

I know that when an object blocks light a shadow is formed. ◯

I can change the size of a shadow. ◯

Light and Dark

17

2 Electricity

In this module you will:
- find out what makes simple circuits work
- understand how a switch can be used to break a circuit.

💬 What is electricity?

Not many people understand electricity but we use it all the time.

💬 We use electricity from the minute we wake up to the second we go to sleep. Can you think of any examples?

Word Cloud

bulb
guess (predict)
circuit
switch   battery
wire

✏️ Think of all the things you have used today that need electricity. Name three.

✏️ If there was no electricity, what would you miss?

Amazing fact

Over one billion people do not have electricity in their homes! Can you imagine that?

Simple circuits

Find out what things make an electric circuit.

The Big Idea

Batteries make things work.

💬 Batteries store electricity. Can you think of any objects that use batteries?

You might have included things like torches and toys in your list. We use lots of different types of battery. Cars and buses use big batteries but watches use tiny batteries.

If you have a toy that needs lots of electricity then a bigger battery is used. Sometimes batteries need to be very small so that they can fit into little spaces like in a hearing aid.

✏️ Look at the photographs below. Fill in the labels to show how the batteries will be used. Use the words in the word bank below to help you.

Word Bank

Car   Torch   Toy car   Watch

Amazing fact

This is the biggest battery in the world. It is bigger than a football pitch. It can make enough electricity for 12 000 homes for 1 hour!

💬 Where else can you get electricity from?

Lots of things use mains electricity. TVs use mains electricity. So do washing machines. People plug them into a socket on the wall.

💬 Why do we sometimes use batteries?

⚠️ Mains electricity is dangerous. Do not put anything into an electrical socket, even if you think it is switched off.

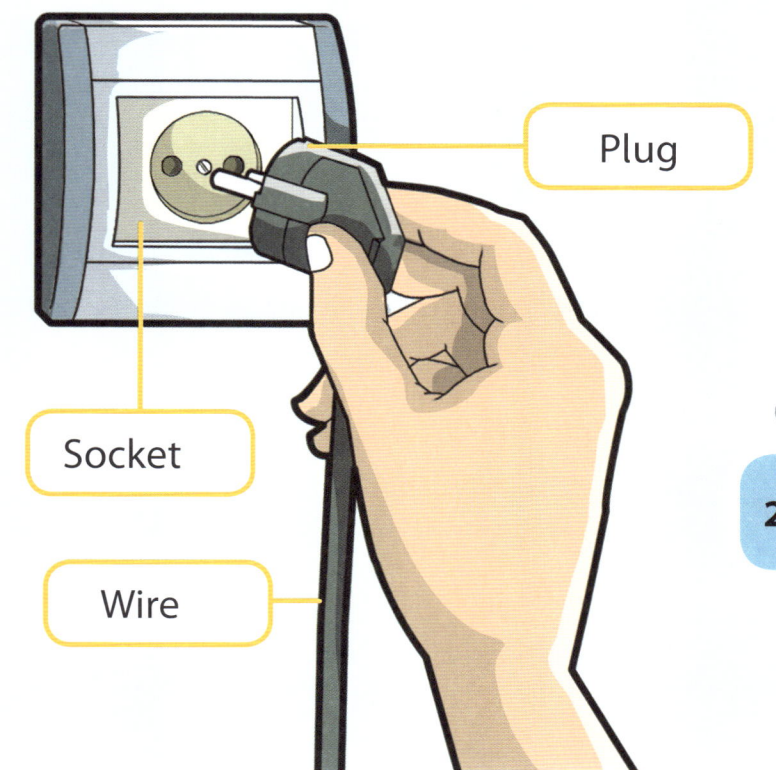

Plug

Socket

Wire

Simple circuits

Find out what things make an electric circuit.

The Big Idea

Wires carry electricity.

💬 Why do we need **wires** for electricity?

Wires carry the electricity from a **battery** or the mains to all the different things that use electricity.

✏️ We call the things that use electricity appliances. Can you name three appliances?

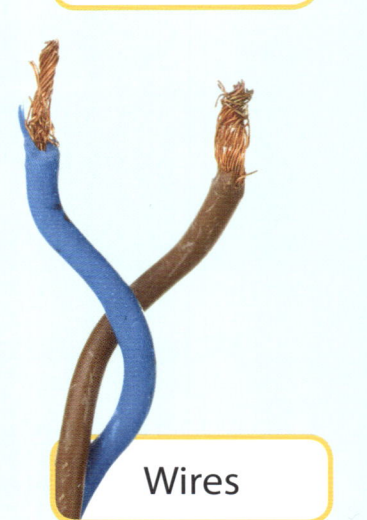

Battery

Wires

✋ Investigation: Building a circuit

1. Connect the **bulb**, wires and battery together to make a **circuit**.
2. Make sure everything is properly connected.
3. What happens to the bulb?

If the bulb lights up, all the parts of the circuit are working and you have done everything correctly. This kind of circuit is known as a simple circuit.

It is called a circuit because everything has to be joined together in a circle for the parts to work. The parts that fit into a circuit are called components.

Bulb

Some materials do not allow electricity to flow through them but others do. Wires are made from materials that let electricity pass through them easily. We can investigate which materials let electricity run through them by building a circuit.

✋ **Investigation: Investigating different materials**

1 Make a circuit three times. Use a different kind of material each time. Predict which material you think will conduct electricity.

2 Test each material, one at a time.

✏️ Name the materials that lit the bulb.

If the bulb lights up, this means the material would make a good wire. It is letting electricity flow through into the other wire. If it does not light up, then the material would not make a good wire. It is not letting electricity through.

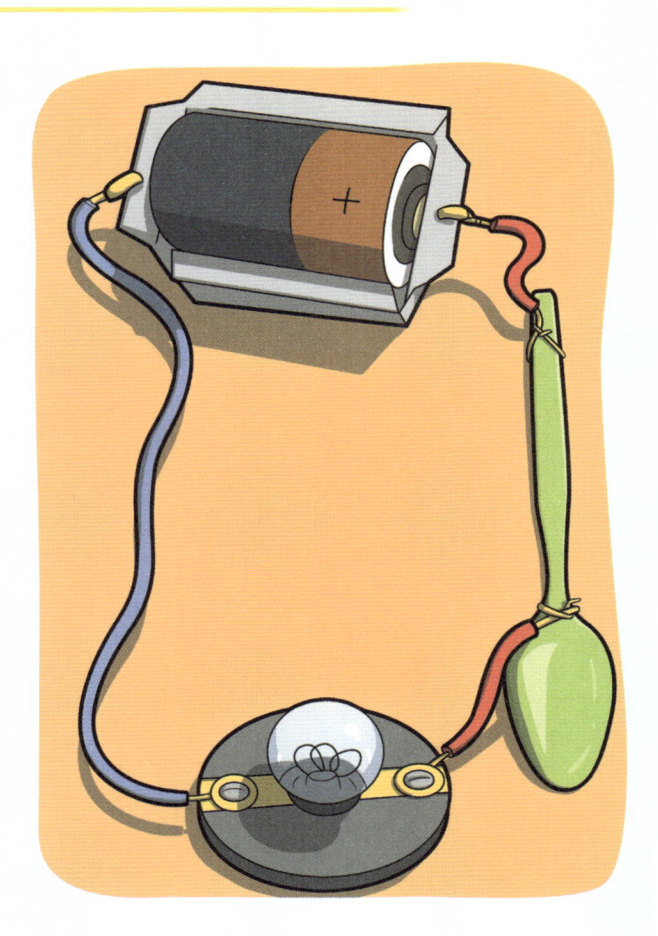

Simple circuits

Find out what things make an electric circuit.

The Big Idea

Buzz!

✏️ List all the components you have used so far to make a circuit.

> **Remember**
> A component is anything that fits into a circuit.

Electricity is not just used to give us light. There are other uses too. So sometimes you need other components in a circuit. A buzzer can be one of them.

✋ **Investigation: Making a buzzer work**

1 Connect all the parts of the circuit together.
2 Does the buzzer buzz?

If the buzzer does not buzz, there must be a fault in the circuit. To find out if the fault is the buzzer, the bulb, the battery or the connections, you need to set up a test circuit. There is a picture of a test circuit opposite.

✏️ Test the connections or components in your circuit. Where was the fault?

A simple circuit contains a bulb, a battery and connecting wires

- Battery
- Wire
- Bulb
- Wire

A test circuit

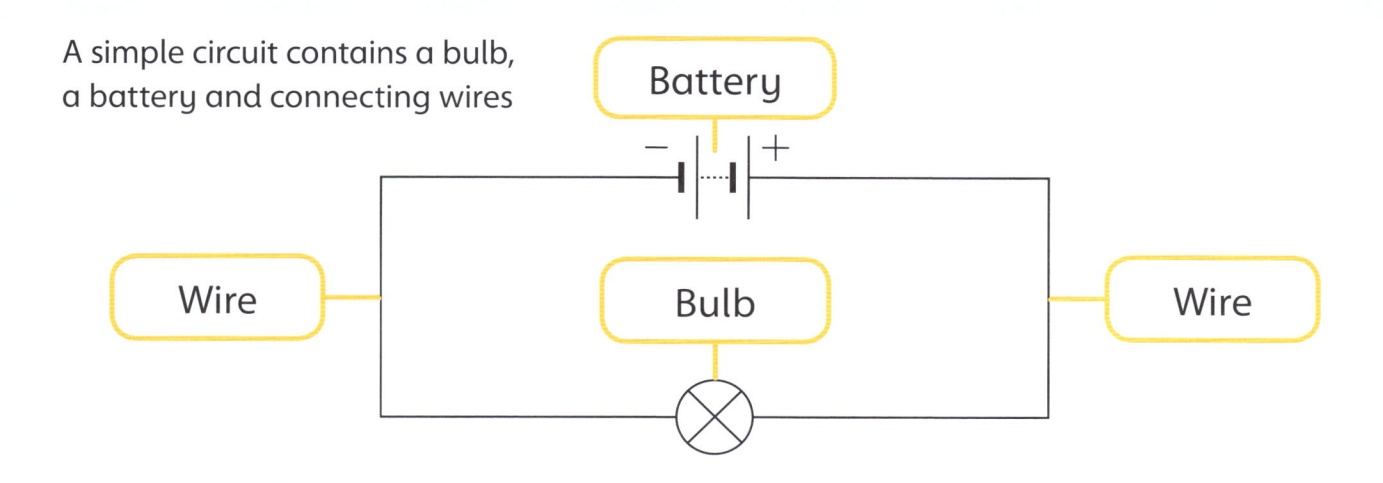

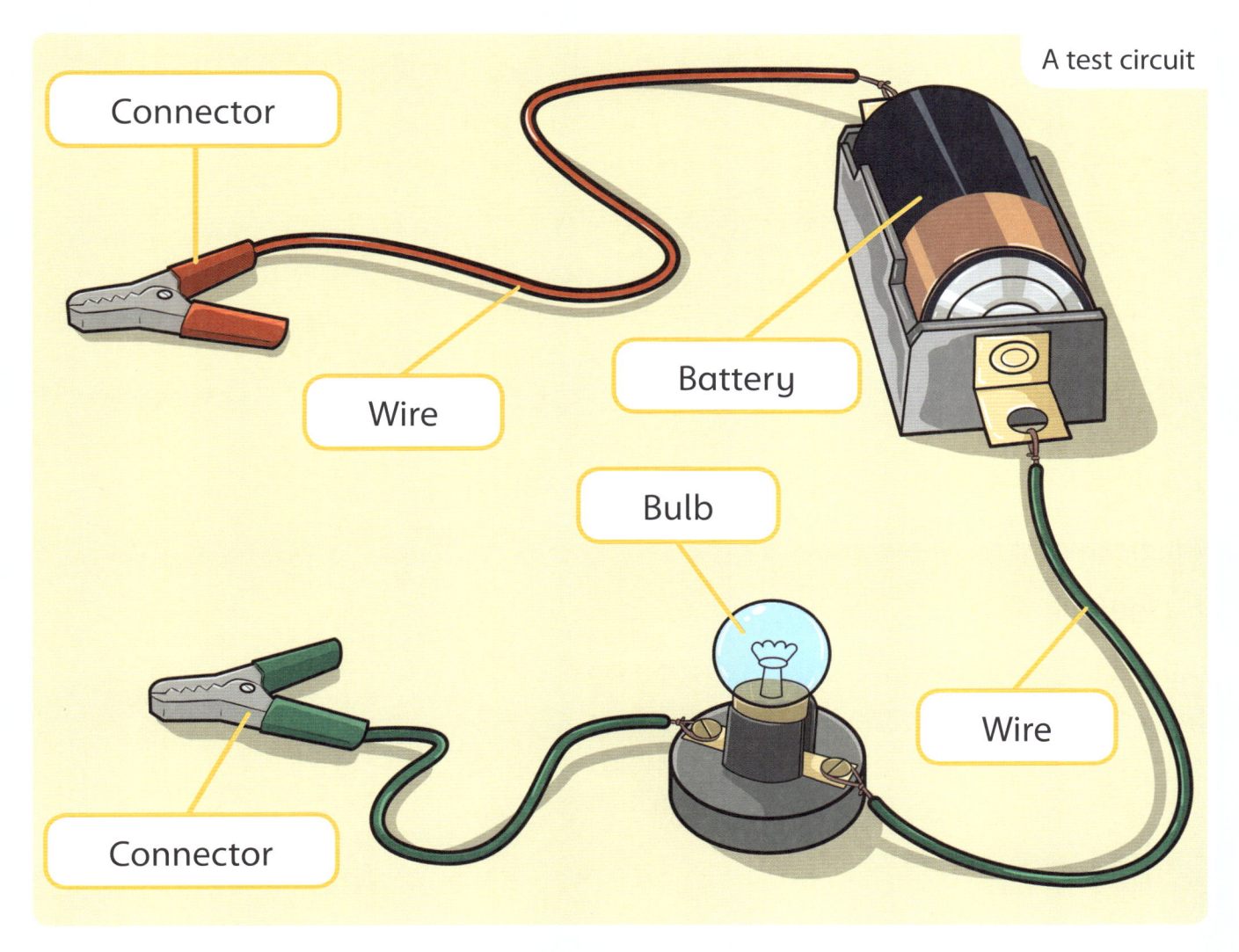

- Connector
- Wire
- Battery
- Bulb
- Wire
- Connector

Q Why do you think it might be useful to have a buzzer in a circuit?

Electricity

25

Simple circuits

Find out what things make an electric circuit.

The Big Idea

We need to be safe when using electricity.

Electricity can be used to make us safe. When a buzzer is added to a circuit it makes a noise, which attracts our attention.

💬 When do we want to make a loud sound like this?

Buzzers are used in lots of different places. If you use a doorbell or car horn you can get someone's attention. Sirens in emergency vehicles are also buzzers in action.

💬 Why do you think modern smoke alarms are connected to the mains?

💬 Why do they also have a battery?

Smoke alarms save at least one life every day by using buzzers.

Amazing fact

Emergency vehicles, such as ambulances and police cars, used to have bells instead of buzzers, but buzzers are much louder than bells. Smoke alarms use a buzzer to attract attention when it comes into contact with smoke. This saves lives.

Mains electricity can be very dangerous if we do not take care.

> ⚠ It is very dangerous to touch a broken wire or socket. This can cause electric shocks that can make us very ill or burn our skin. Electricity should also never touch water.

An electric shock is caused when electricity goes into you. This is very dangerous. When using electricity we should be very careful.

✏ Why should you always dry your hands before touching a light switch?

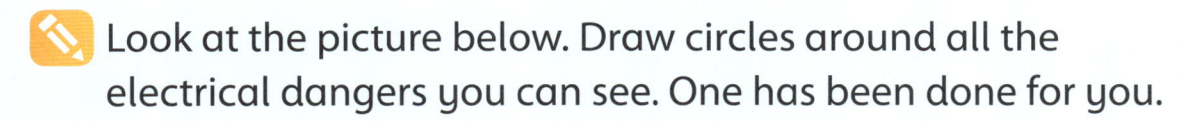

✏ Look at the picture below. Draw circles around all the electrical dangers you can see. One has been done for you.

Electricity

27

Simple circuits

Find out what things make an electric circuit.

The Big Idea

Electricity can be used in games.

✏️ Make a list of the games you have that need electricity to work.

✋ Investigation: Battery matching

1 Look at the appliances your teacher gives you. Choose one to investigate.
2 Guess how many batteries it will need.
3 Choose your batteries and put them in.

💬 Did you notice + and − on any of the batteries and battery compartments? Why do you think they are there?

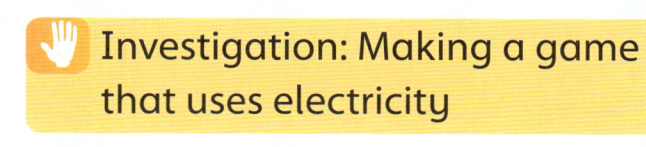

✋ **Investigation: Making a game that uses electricity**

1 Test your components.
2 Connect all your components together to make the steady-hand game. Make sure you can hear a buzz!
3 Play your game!

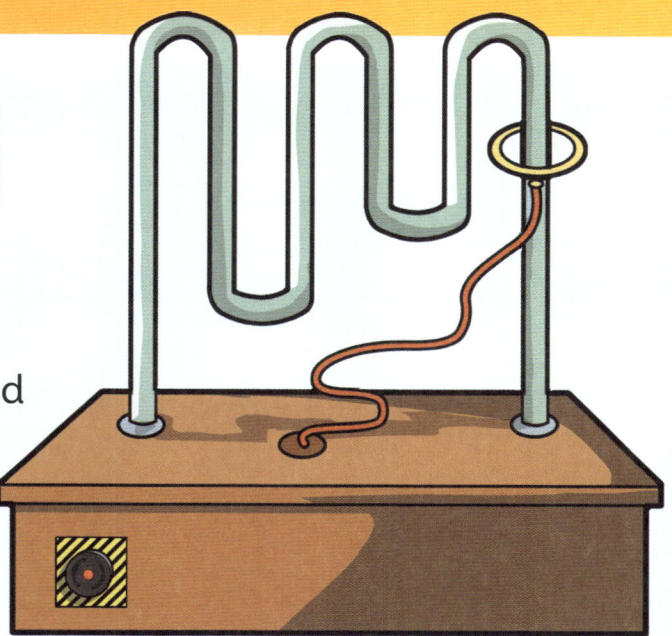

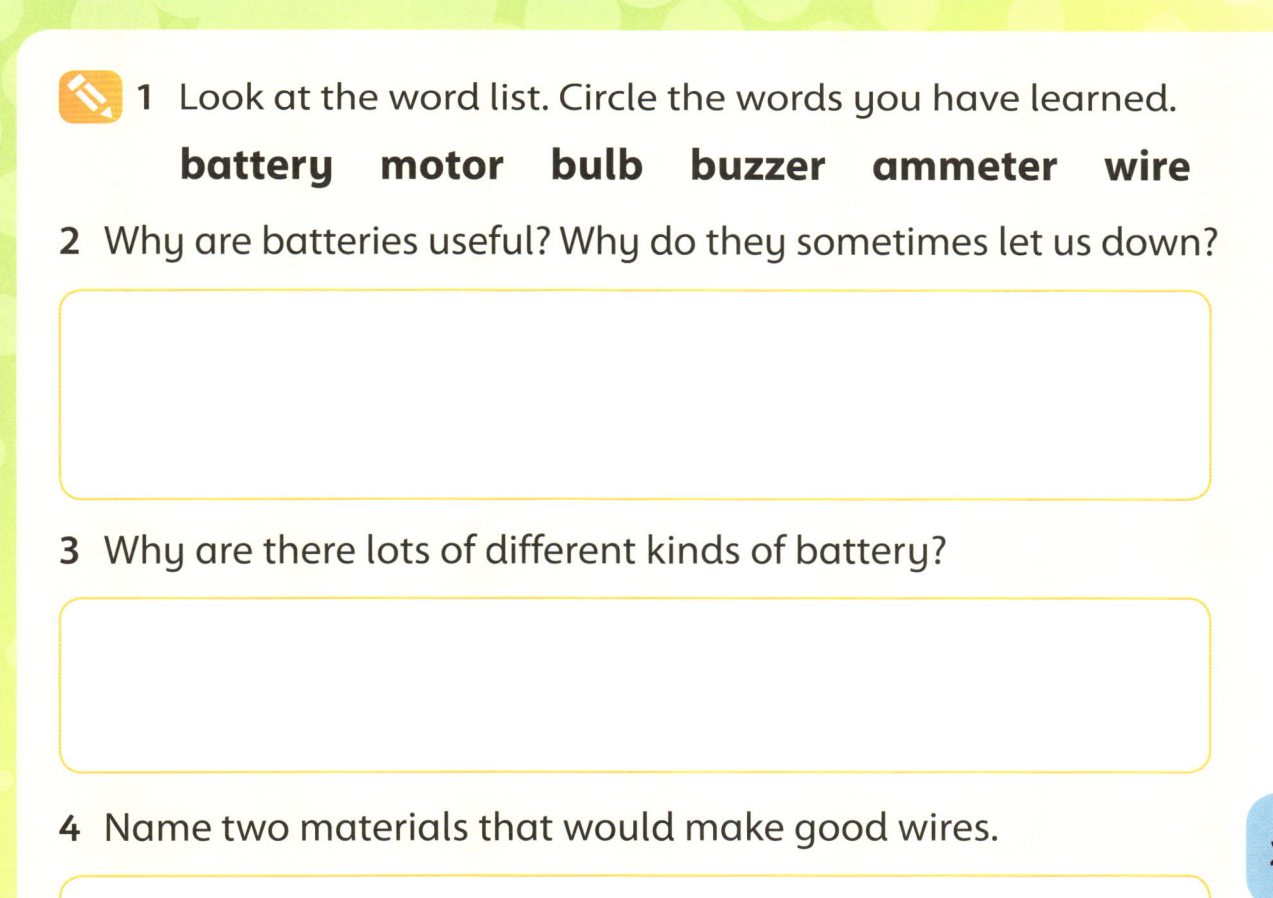

1 Look at the word list. Circle the words you have learned.

battery   motor   bulb   buzzer   ammeter   wire

2 Why are batteries useful? Why do they sometimes let us down?

3 Why are there lots of different kinds of battery?

4 Name two materials that would make good wires.

Now turn to page 36 to review what you know.

Designing, constructing and predicting

Learn how a switch can break a circuit.

The Big Idea

We can stop and start electricity.

💬 Why do we sometimes need to stop electricity?

💬 What happens to a circuit if there is a gap in it?

Look at the first picture. Can you see that the **switch** is being turned off? This makes a gap in the circuit. This means that electricity cannot flow to the device and it does not work.

Now the switch has been turned on. This means that the circuit has been completed and electricity flows to the device.

Look at the picture below of a circuit with an open switch. There is a gap between the two pieces of metal in the switch.

✏️ Is the switch in the picture on or off?

Open switch

Look carefully at this picture of a closed switch. There isn't a gap between the two pieces of metal. This means the circuit is complete and electricity will flow.

Closed switch

✏️ Is the switch in this picture on or off?

✋ Investigation: Exploring how a switch works

1. Make a simple circuit with a switch in it.
2. Press the switch on and off. Predict what will happen.

Remember
Do you remember how to make a simple circuit? Let's see.

You can make a switch out of lots of things. The material you use must allow electricity to flow through it.

✏️ Which materials let electricity flow through them best?

✋ Investigation: Making a switch

1. Take the switch out of your circuit. Put a paper clip in its place.
2. Slide the paper clip over to the loose crocodile clip.
3. What happens?

The bulb will still light.

💬 Why does the paper clip still light the bulb?

Electricity

31

Designing, constructing and predicting

Learn how a switch can break a circuit.

The Big Idea

Guess which circuit will work.

💬 Why is a circuit called a circuit?

The word 'circuit' comes from the word 'circle'. Electricity can only flow around a circuit if all the components are connected and there are no breaks in the wires.

💬 Do you know what the word 'predict' means?

✏️ Look at this picture of a circuit. Will the bulb light or not?

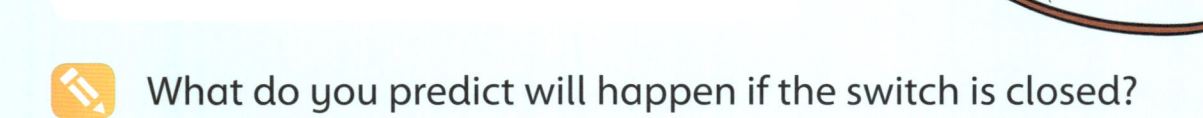

✏️ What do you predict will happen if the switch is closed?

✏️ If this circuit did not work even with the switch closed, what could we do to find out where the problem is?

✋ **Investigation: Exploring circuits**

1 Look at these pictures of circuits. All the bulbs are shown as lit. Is this right? Which do you predict will work?

a   b   c

2 See if you are right by making each circuit.

3 Was your prediction right?

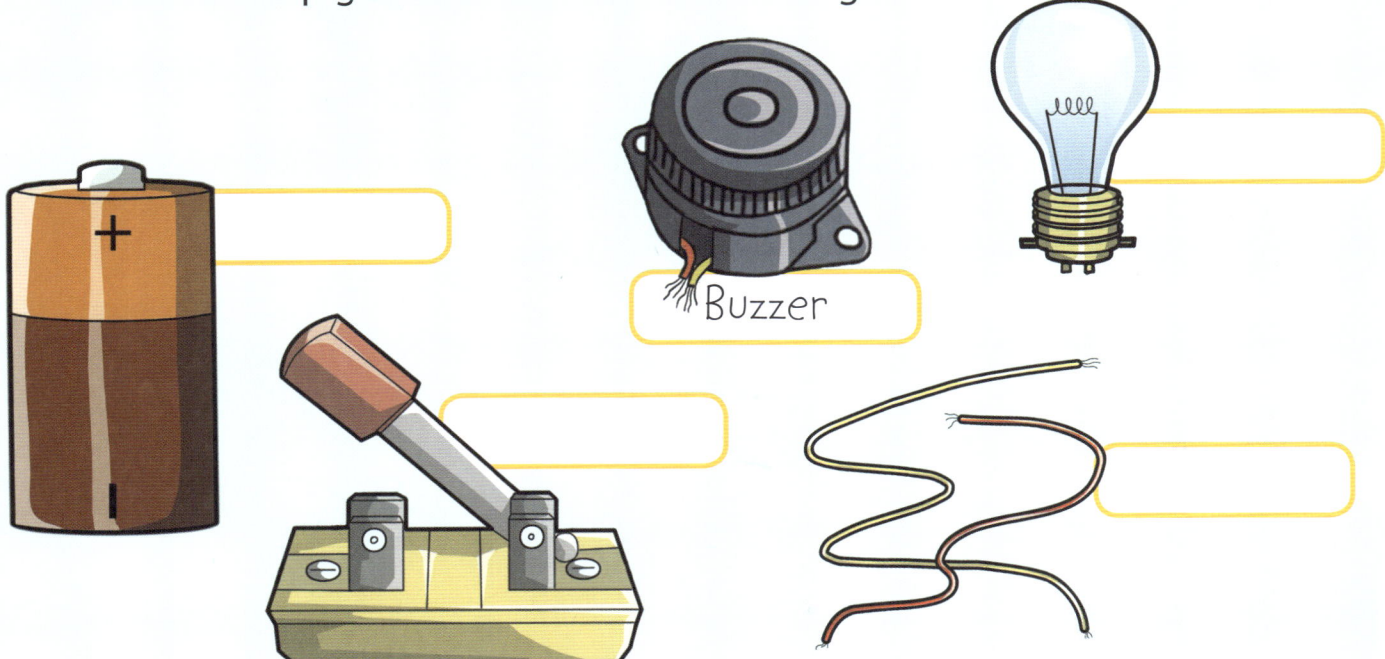

✏️ Fill in the labels for the component names. Use the word bank below to help you. One has been done for you.

Buzzer

Word Bank

Battery   Bulb   Switch   Wires   ~~Buzzer~~

33

Electricity

Designing, constructing and predicting

Learn how a switch can break a circuit.

The Big Idea

Wiring a model house.

✏️ Look around your classroom. Make a list of all the light switches and lights in the room.

Look at the picture of the model house. We are going to make working lights for it. Your teacher will give you a room to think about.

✏️ Where will you put your lights and light switches in your room? Draw a picture of your room and label it to show where the lights and switches will go.

💬 Can you add a buzzer as a smoke alarm or doorbell?

✏️ **1** Are switches circuit breakers?

2 Why are two or more batteries sometimes needed?

3 Why shouldn't you use a broken wire?

Now turn to page 37 to review what you know.

What we have learned about electricity

Simple circuits (pages 20–9)

✏️ Give two reasons why you would use a battery, not mains electricity, to power a toy.

✏️ Why do we have to be very careful with mains electricity?

✏️ Name three components of a circuit.

✏️ Why is it called a circuit?

I know at least three components of a circuit. ◯

I understand why we do not use mains electricity for our toys. ◯

36

Designing, constructing and predicting
(pages 30–5)

✏️ What happens when a circuit is broken?

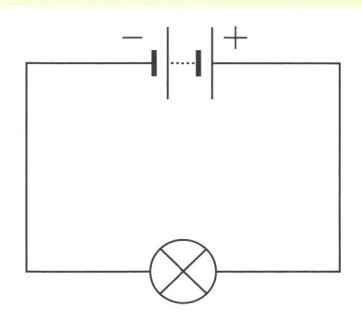

✏️ Which component do we use to break a circuit?

✏️ If your circuit did not work, how could you find out why?

✏️ Would you need more than one battery to light lots of bulbs?

I know why a broken circuit does not work. ◯

I understand how a switch works. ◯

I can build a switch. ◯

Electricity

37

3 Changing Materials

In this module you will:
- look at the ways we can change materials
- sort different materials into groups
- learn that some materials are found naturally and some are made.

Word Cloud

material
melt
bend
solid
natural
stretch
twist

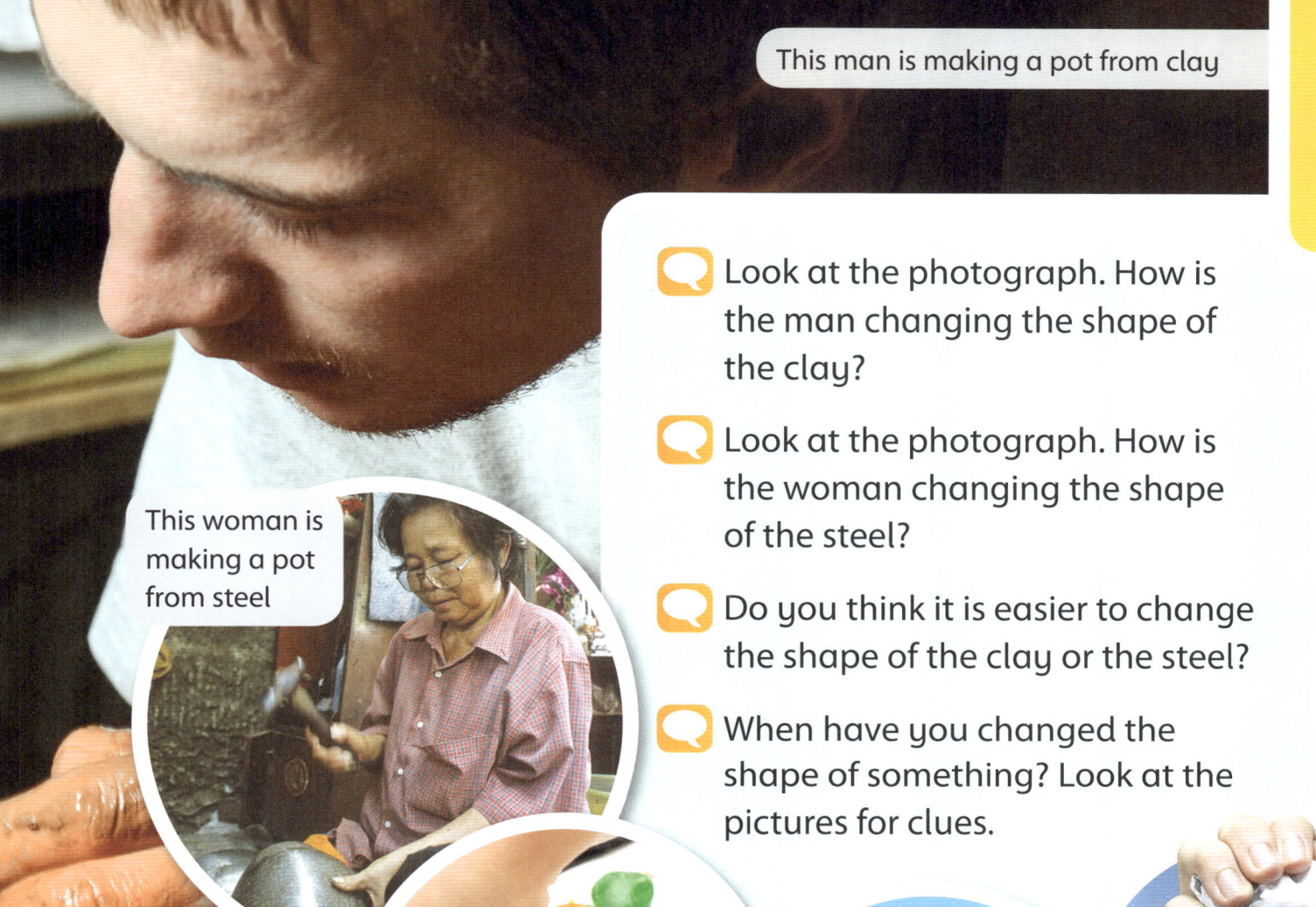

Chemistry

This man is making a pot from clay

This woman is making a pot from steel

- Look at the photograph. How is the man changing the shape of the clay?

- Look at the photograph. How is the woman changing the shape of the steel?

- Do you think it is easier to change the shape of the clay or the steel?

- When have you changed the shape of something? Look at the pictures for clues.

Changing Materials

39

Amazing fact

Changing the shape of materials isn't a new idea. People who lived more than 10 000 years ago made metal jewellery.

Changing the shapes of materials

Understand that the shapes of some materials can be changed.

The Big Idea

We can make things change shape by squashing, bending and stretching them.

💬 Think about what you have done today. How many ways have you changed the shape of something?

✋ Investigation: Making a paper picture

1. Draw a picture in pencil on a large piece of card.
2. Take some pieces of coloured tissue paper and scrunch them up into balls.
3. Stick the coloured balls to your picture.

✏️ How did you change the shape of the paper?

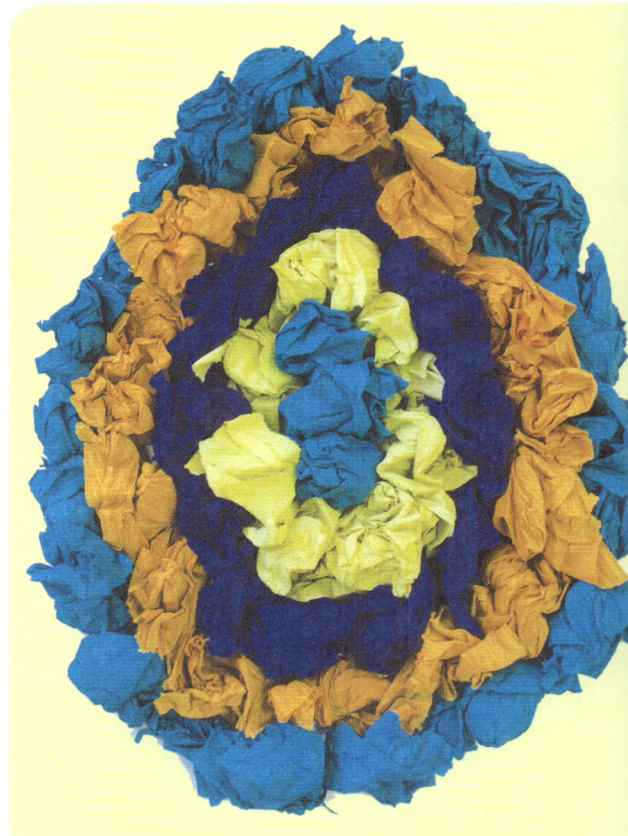

Bending, squashing and twisting a **material** will change its shape. Another way to change the shape of a material is to **stretch** it. Let's find out which materials stretch the most!

✋ **Investigation: Stretching materials**

1 Measure your material.
2 Fix your material to the holder.
3 Hang the weights from it with a paper clip.
4 Measure how far the material has stretched.

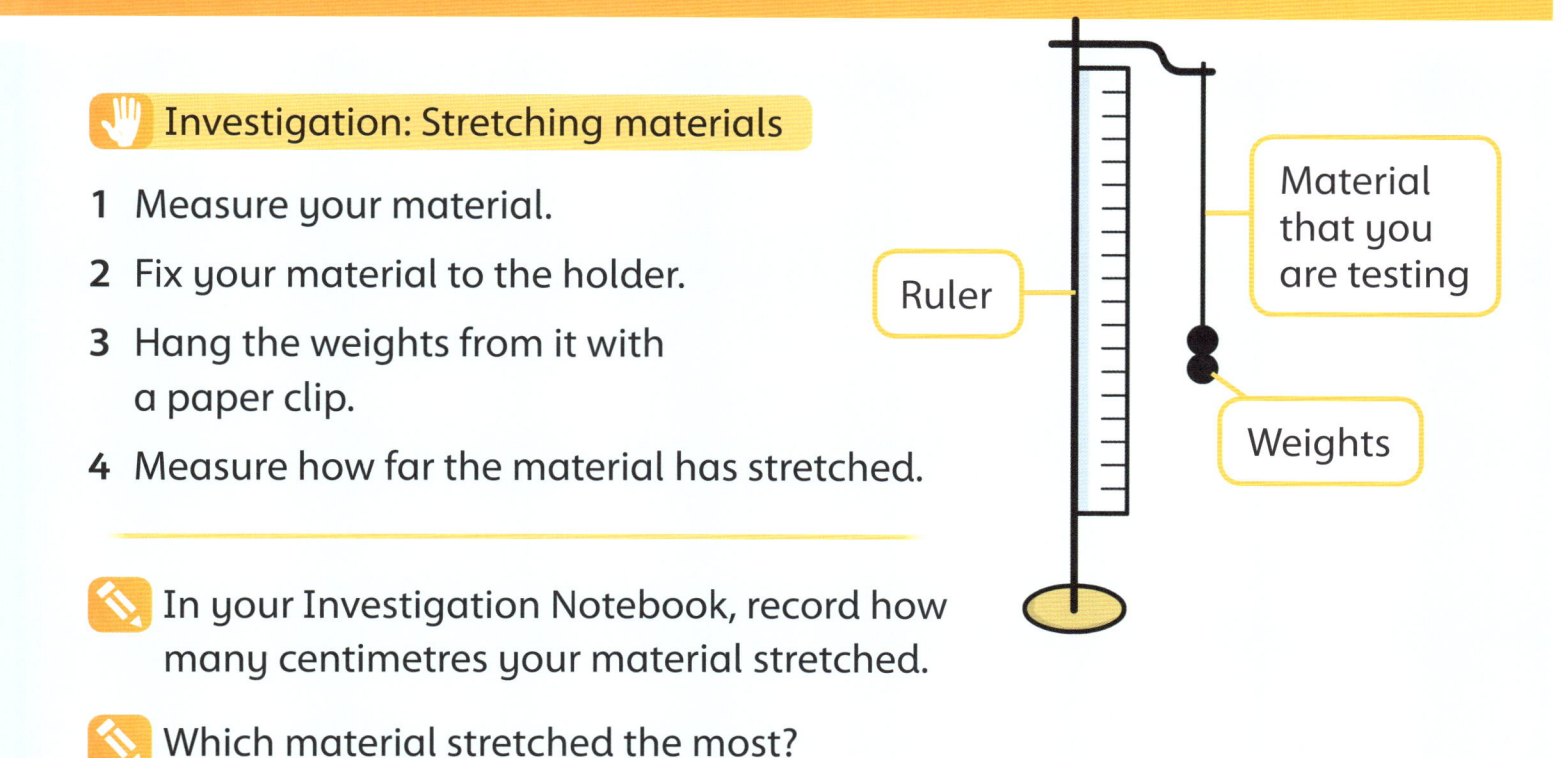

Ruler

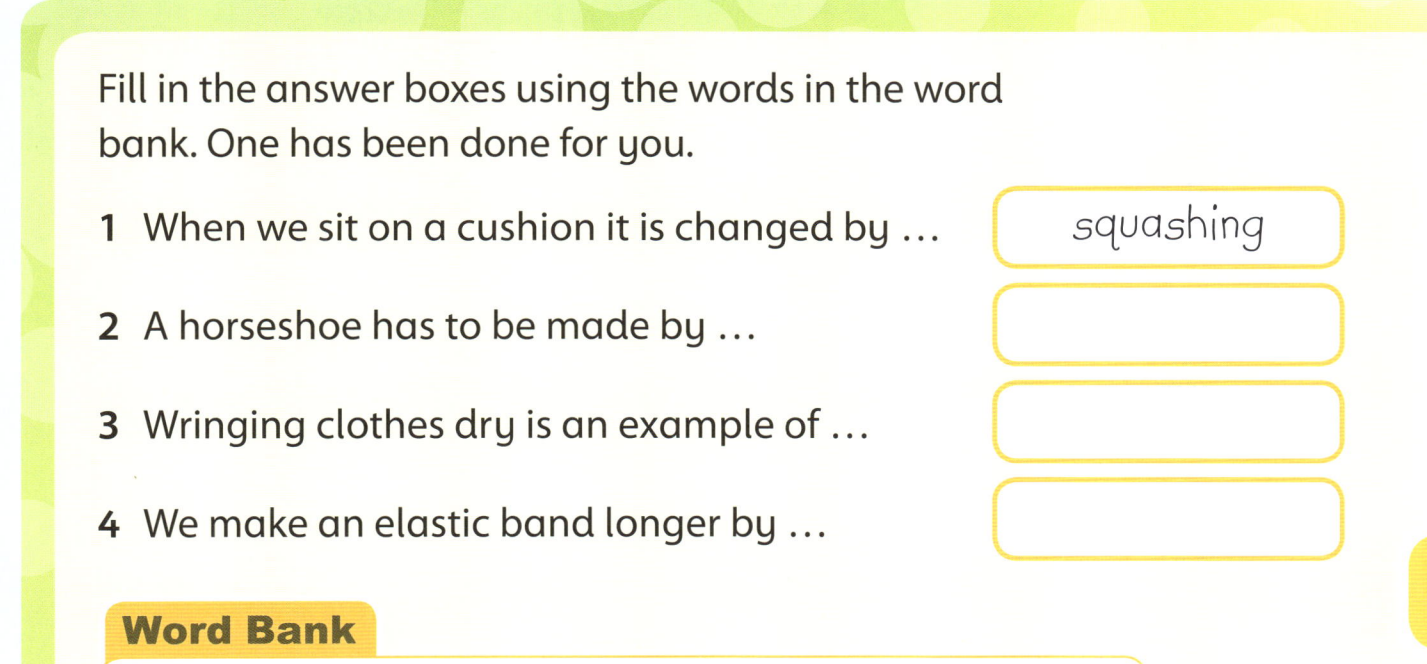

Material that you are testing

Weights

📝 In your Investigation Notebook, record how many centimetres your material stretched.

📝 Which material stretched the most?

Fill in the answer boxes using the words in the word bank. One has been done for you.

1 When we sit on a cushion it is changed by … | squashing

2 A horseshoe has to be made by … |

3 Wringing clothes dry is an example of … |

4 We make an elastic band longer by … |

Word Bank

~~squashing~~   bending   twisting   stretching

Changing Materials

41

Now turn to page 52 to review what you know.

Heating materials

Look at the way some everyday materials change when they are heated or cooled.

The Big Idea

We can change things by heating them.

✏️ List two ways that you changed materials in the last lesson.

💬 Look at the photographs. Do you know what they show?

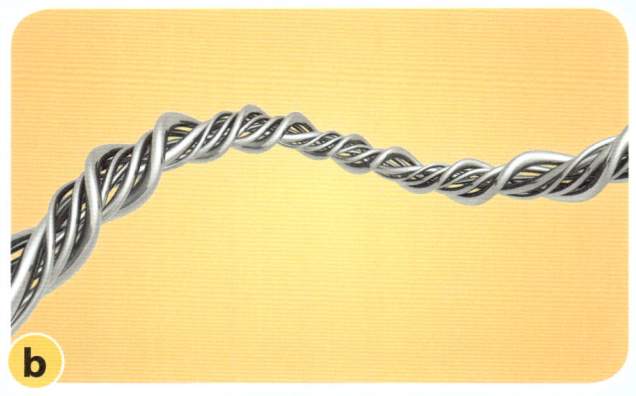

a

b

The shape of these metals has been changed. One has been changed by **bending**. The other by **twisting**.

💬 Which one has been changed by bending? Which one has been changed by twisting?

Materials can also be changed by heating or cooling them. Heating something means to make it hotter. Cooling something means to make it colder.

✋ **Investigation: Making clay bowls**

1 Split your clay into two equal pieces.

2 Make each piece into a small bowl of the same size and shape.

3 Put one bowl into a cool, damp place. Your teacher will heat the other one so that it gets very hot.

💬 What do you think will happen to the two bowls?

💬 Now that you have your bowls back, can you see any differences between them?

💬 Were your predictions correct?

The heat has made one of the clay bowls much more useful. It is not as soft as the clay you started with. We cannot use the bowl if it is soft.

Changing Materials

43

Heating materials

Look at the way some everyday materials change when they are heated or cooled.

The Big Idea

We can change things by heating them.

Lots of other materials change when they are heated.

Investigation: How foods change when they are heated

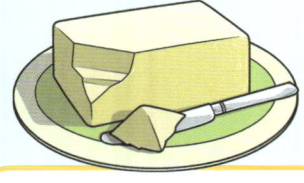

125g butter

125g caster sugar

2 medium eggs

2 tablespoons of milk

125g self-raising flour

1. Mix together the butter and sugar in a bowl until the mixture feels light and fluffy.

2. Add the eggs, flour and milk to the bowl and mix it all together until the mixture feels smooth.

3. Put your mixture into paper cases and bake the cakes in the centre of the oven.

⚠️ Your teacher will put the cakes in the oven, then take them out when they are ready and put them somewhere to cool.

✏️ Why does your teacher have to be careful when putting the cakes in and taking them out of the oven?

✏️ Do the baked cakes look like any of the ingredients?

✏️ How do the cakes look different from the ingredients?

✏️ How do the cakes feel different?

Amazing fact

Making steel is a bit like baking a cake. Ingredients such as iron, other metals and carbon are added together and heated. Your cake was baked at 200°C. The steel is 'baked' at 1200°C!

Now turn to page 52 to review what you know.

Cooling materials

Look at the way some everyday materials change when they are heated or cooled.

The Big Idea

We can change things by cooling them.

We know that materials can change when they are heated. Materials can also change when they are cooled.

Look at the photograph of the ice cream. Some materials change a lot when they are cooled. Cream is a liquid. When cream is cooled it changes from runny to hard. It has changed into a **solid**.

Investigation: Cooling water

1 Fill your plastic containers with water.
2 Put the containers into a freezer.
3 Look at your containers carefully when you have them back.

What do you see and feel?

The water you started with was a liquid. When liquid water is cooled it can turn into a solid. This solid is called ice.

When ice turns back to water it is called **melting**. The solid ice melts to give liquid water.

✏️ What would you do to get liquid water from ice?

✋ Investigation: Heating ice

1 Break some of your ice into big pieces. Break the rest up into small pieces.

2 Put half of your big pieces and half of your small pieces in a warm place. Put the rest in a cool place.

3 Look at your ice samples every ten minutes. What do you see?

✏️ Fill in the missing words. Use the words in the word bank. One has been done for you.

Materials can be changed by heating and _____. Baking a _____ is an example of heating materials. Water can cool down and make ___ice___. This can be heated to make water again. If we carry on heating the water it will turn into _____. This can be cooled and it will turn back to _____.

Word Bank

cake   steam   cooling   water   ~~ice~~

Now turn to pages 52–3 to review what you know.

Dissolving materials in water

Understand that some materials can dissolve in water.

The Big Idea

Dissolving materials in water is very useful.

💬 Look at the photograph of a cup of tea. Do you know how a cup of tea is made?

Not all materials will spread out in water. If a material spreads out in water so that we cannot see it we say it has dissolved. The material may colour the water but we cannot see the pieces of the material. If a material does not spread out in water we say it has not dissolved.

Look at the pictures below. They show materials before and after they have been put into water.

✏️ Complete the labels to show which material has dissolved and which has not dissolved.

48

✋ **Investigation: Dissolving**

1. Add some water to your beakers so they are all half full.
2. Predict whether the materials will dissolve or not.
3. Add a spoonful of the first material to your first beaker. Stir.
4. What happens? Was your prediction right?

✏️ Which substances dissolved? Which substances did not dissolve?

💬 Look at the photograph of the sea. Where does the salt go when it dissolves?

We know that sea water is salty but the sand on the beach does not dissolve into the sea.

💬 How could you show that when things dissolve they do not disappear?

✏️ Circle the correct answer.

1. When a material spreads out in water we say it has dissolved. True   False
2. Only some materials will dissolve in water. True   False
3. Sand dissolves in water. True   False

Now turn to page 53 to review what you know.

Natural or not natural?

Understand that some materials occur naturally and others are man-made.

The Big Idea

We use materials that are found in nature but also some that are made by people.

A display of natural materials

Some materials have been used for making things for thousands of years.

✏️ Match the material with how it is used. One has been done for you.

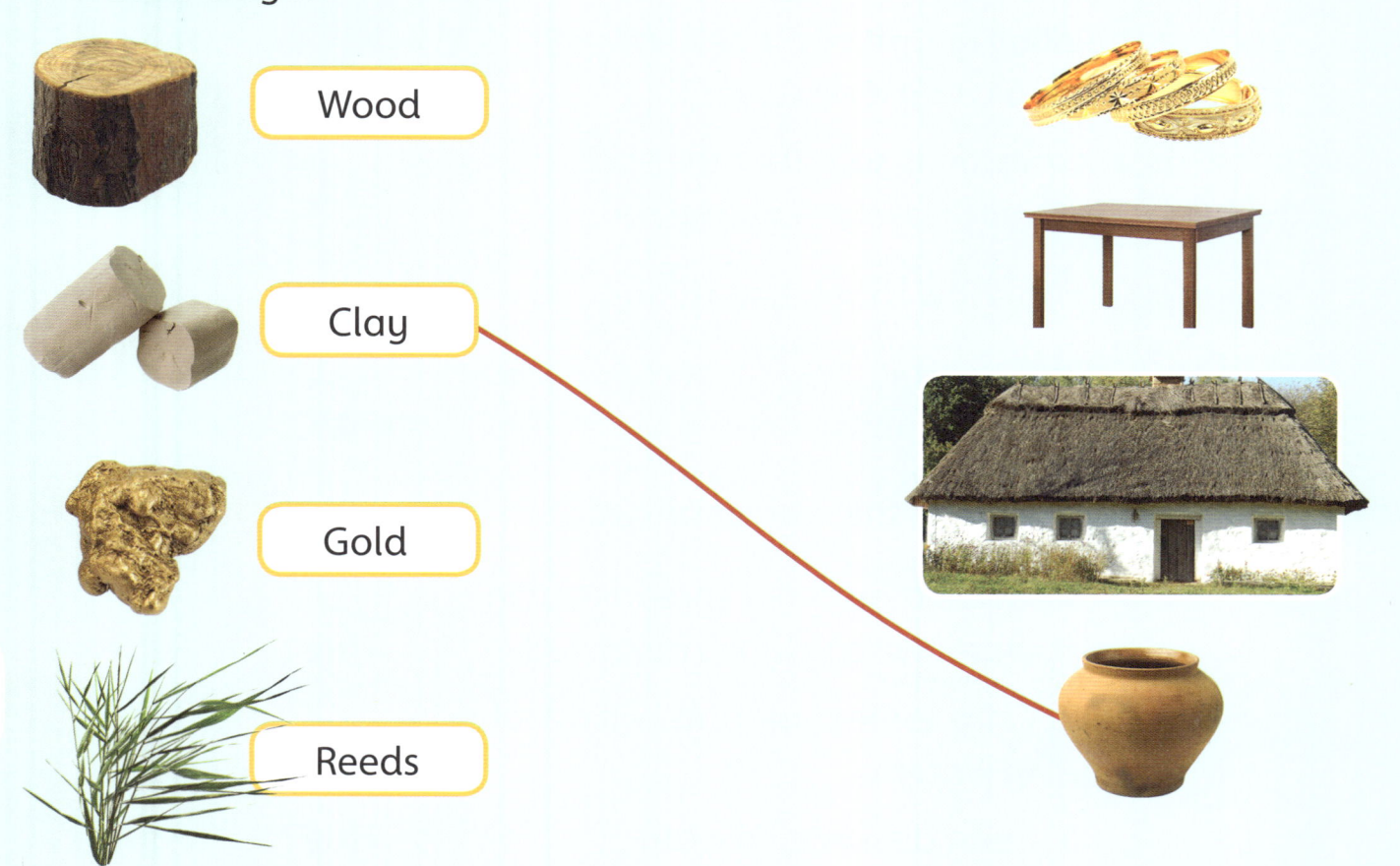

- Wood
- Clay
- Gold
- Reeds

50

Materials that are found in nature are called **natural** materials, such as wood, clay, stone, shell and sand.

Materials that people make are called man-made or not natural, such as glass, plastic, concrete and steel. Many buildings, roads, dams and bridges are made from concrete.

✋ Investigation: Sorting objects – natural or man-made?

1. Look at the objects your teacher has given you.
2. Decide which materials the objects are made from.

Think about...
What would your school look like if there were no man-made materials? What would you miss?

1. Circle three natural materials. Underline three man-made materials.

| wood | glass | plastic | stone | steel | stone |

2. Which natural material is often used for making pots and bowls?

3. Which man-made material is used to build bridges, dams and roads?

Now turn to page 53 to review what you know.

What we have learned about changing materials

Changing the shapes of materials (pages 40–1)

✏️ Name three ways that you can change the shape of materials.

I know four ways to change the shape of materials. ◯

Heating materials (pages 42–5)

✏️ How could you change the shape of a hard metal or glass?

✏️ How could you make clay or bread dough hard?

I can name two things you can change the shape of by heating them. ◯

Cooling materials (pages 46–7)

✏️ Name two things in the kitchen that need to be kept cool to keep their shape.

I know that cooling some things can make them hard. ○

Dissolving materials in water (pages 48–9)

✏️ Name three things that dissolve in water.

✏️ Name three things that do not dissolve in water.

I know that not all things can dissolve in water. ○

Natural or not natural? (pages 50–1)

✏️ Circle the correct answer.

1 Concrete is …   natural   man–made
2 Wood is …   natural   man–made

I understand that when people change a material into a different material it becomes man-made. ○

Changing Materials

53

4 Looking at Rocks

In this module you will:
- look at different types of rock
- find out about what rocks are used for
- investigate different types of soil.

💬 Look at the photographs. Do you know what the rock formation and the building are made from?

Word Cloud: rock, sandy, stone, soil, compare

Amazing fact
The oldest rocks on Earth are nearly four billion years old.

Chemistry

💬 Look at the rocks your teacher has given you. How do they feel?

💬 Look at the rock formation and building again. How do you think they would feel if you touched them?

💬 Look at the pictures. Which of the building materials have been made by people? Which can be found in nature?

- Wooden rafters
- Metal girders
- Glass windows
- Plastic window frames
- Building bricks
- Building stones

Looking at Rocks

55

What are rocks?

Recognise different rocks.

Understand that rocks can be used in different ways.

The Big Idea

Rocks are very important and are found everywhere.

Look at the photographs. They show how **rocks** are used today and how they were used in the past.

- Where have you seen rocks used in buildings? Have you seen any in your school?
- Why is rock such a useful material for buildings?
- Look at the picture of the beach. Circle any examples of rock you can see.
- What are stones?

The **stones** you find on the ground are broken-down pieces of rock. Round stones are called pebbles.

Sand is also a type of rock. It is made from tiny grains of rock that have been broken down by the weather.

✏️ Look around your school. List the parts of the building made from stone.

Sand and pebbles can be found on the beach and in deserts

Rocks are found everywhere and we can see them all around us. They are also under the ground and at the bottom of the sea.

✏️ Look back at the picture of the beach. Do you want to circle any more examples of rock?

Think about...
What is under the school? If you dug down deep enough, what might you find?

Now turn to page 68 to review what you know.

Different building materials

Recognise different rocks.
Understand that rocks can be used in different ways.

The Big Idea

We use many different building materials to make our modern world.

💬 The photograph shows an ancient temple. What is the temple made from?

Rocks have been used for buildings for thousands of years. The rock is often cut into smooth blocks. Rock has also been used for making roads and bridges, and for carrying water from one place to another.

These rocks have been helping to carry water for thousands of years

Look at the photograph. Modern buildings still use rock. This picture from Abu Dhabi shows how rocks can be used to make beautiful buildings today.

Rock is found in nature. We call it a natural building material. Other natural building materials are wood, clay, straw and bamboo.

✋ **Investigation: Making bricks**

Clay can also be used to make building bricks.

1. Take some clay and squash it in your hands until it feels nice and soft.
2. Press the clay with your hands until it is the shape of a brick.
3. Leave the clay to dry. It will make a hard brick.

Not all building materials are found in nature. Some are made by people. These are man-made materials like glass, plastic, concrete and metals.

✏️ Look at the pictures. Circle all the man-made building materials.

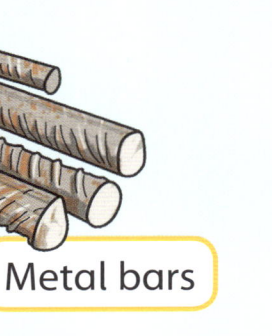

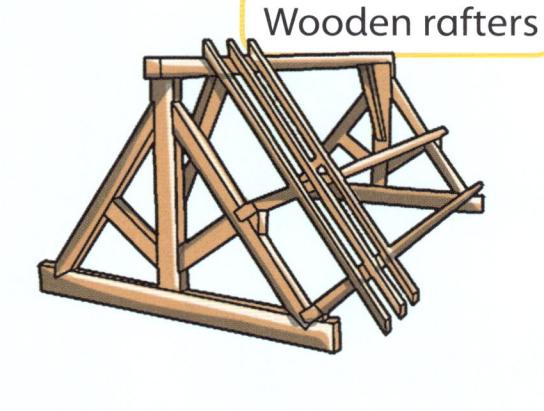

Concrete blocks

Wooden rafters

Glass windows

Metal bars

Clay bricks

Plastic window frames

Now turn to page 68 to review what you know.

Types of rock

Recognise different rocks.

Understand that rocks can be used in different ways.

The Big Idea

All rocks aren't the same.

💬 Can you remember your work on building materials?

✏️ Name two natural building materials.

✏️ Name two man-made building materials.

Rocks are formed in different ways. Some are made when hot rocks from volcanoes cool down. Some are made from small pieces of sand or seashells. These settle to the bottom of the sea and turn into solid rock. This takes millions of years.

Rock made from sand is called sandstone. Some rocks are made up of tiny pieces of mud. This is called mudstone.

a

✏️ Look at the rocks. What types of rock are they? Complete the labels.

Rock made from seashells is called limestone. Sometimes the shells form fossils in the rock. Chalk also has lots of very tiny shells in it.

b

✏️ Look at this photograph of limestone. Draw arrows to show where there are fossils.

Some rocks are made deep underground. It is very hot there and the rocks are in the form of liquid lava. When the lava pours out of the top of the volcano the rocks begin to cool. This makes them turn solid. These rocks are usually very hard.

The red-hot lava cools to make solid rock

✋ **Investigation: Looking at rock samples**

1 Look at the rock samples.

2 Sort the rocks into the same type. Ask yourself the geologist's questions.

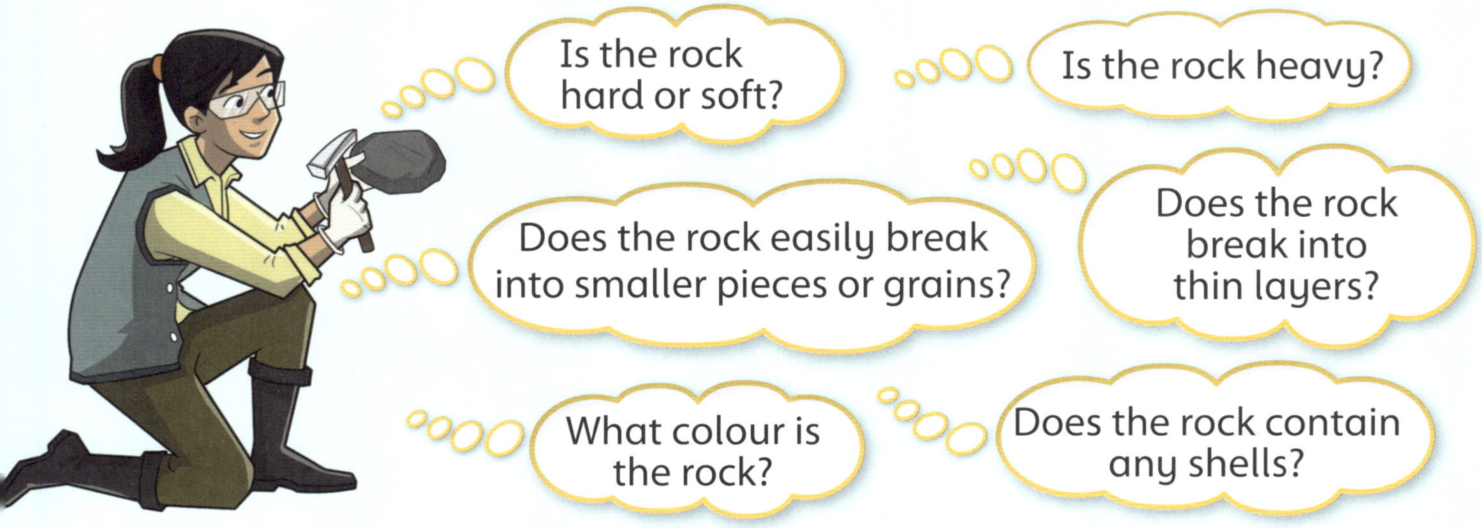

- Is the rock hard or soft?
- Is the rock heavy?
- Does the rock easily break into smaller pieces or grains?
- Does the rock break into thin layers?
- What colour is the rock?
- Does the rock contain any shells?

Now turn to page 69 to review what you know.

How do we use rocks?

Recognise different rocks.

Understand that rocks can be used in different ways.

The Big Idea

Rocks are used for different things.

💬 Do you remember looking at different types of rock?

✏️ Can you remember talking about properties of rocks? Name three properties.

✏️ Which rock had fossils in it?

✏️ Which rock was made up of small grains of sand?

Some of the rocks you tested were very hard. These would be good for building houses and roads. Other rocks look nice. These can be used for the front of buildings. Other rocks are too soft to be used.

Look at the picture. The man is splitting a type of rock called slate.

✏️ Draw a line from the type of rock to how it is used. One has been done for you.

Name and properties

- Sandstone – strong and not easily broken down
- Marble – white, attractive and easily shaped and polished
- Slate – waterproof and breaks easily into sheets
- Granite – very hard, waterproof and attractive
- Chalk – white, soft and wears away easily

How the rock is used

- statues
- steps, fronts of buildings
- walls of buildings
- for writing on blackboards
- roof tiles

(Slate is connected to roof tiles.)

Granite is only found in some places. It is difficult to cut out of the ground, transport and polish. This means it costs a lot of money to buy.

✏️ Why is granite used on the front of a building but not to build all of it?

Now turn to page 69 to review what you know.

Different types of soil

Recognise different rocks.

Understand that rocks can be used in different ways.

The Big Idea

Soil is important for life on Earth. Without it most plants would not grow.

💬 Look at the picture. What are the men doing?

✏️ We eat foods that come from plants. What plants have you eaten this week? Name three.

✏️ What else are plants used for? Give two uses.

Plants need water and they need something to grow in. This is a special mixture called **soil**. Soil provides plants with food, water and support.

Look at the man planting. He is putting small plants into the soil.

Soil has small pieces of rock in it. These can be as small as sand grains or as large as pebbles. Soil has lots of different things in it to help plants grow.

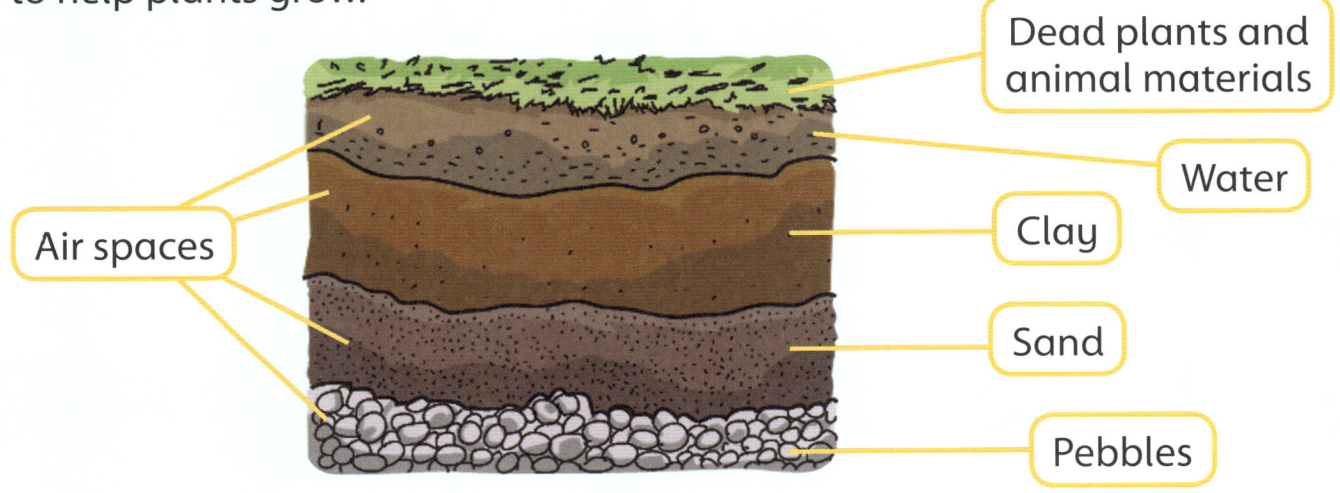

- Dead plants and animal materials
- Water
- Clay
- Sand
- Pebbles
- Air spaces

✋ **Investigation: Finding out what is in a soil sample**

1. Add five spoonfuls of soil to a jar of water.
2. Stir the mixture and then let it settle.
3. What happens? Does your jar look like the picture below?

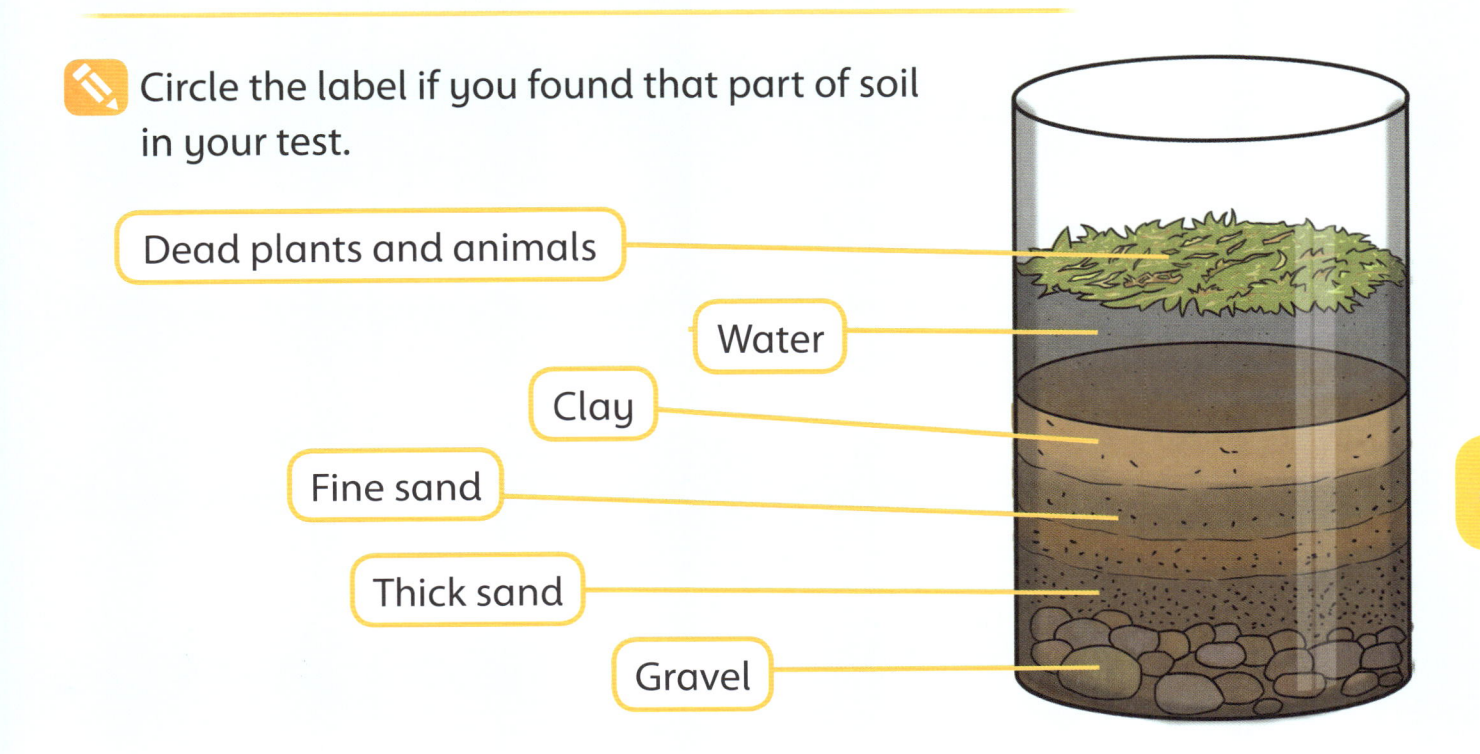

✏️ Circle the label if you found that part of soil in your test.

- Dead plants and animals
- Water
- Clay
- Fine sand
- Thick sand
- Gravel

Looking at Rocks

65

Different types of soil

Recognise different rocks.

Understand that rocks can be used in different ways.

The Big Idea

Soil is important for life on Earth. Without it most plants would not grow.

If water does not run through soil easily it can become too soft and muddy. If water runs through too quickly it can mean plants do not have time to take the water in.

💬 Look at the photographs. Which soil is too wet? Which soil is too dry?

✋ Investigation: How well does your soil drain?

1. Add a small amount of cotton wool to the bottom of a filter funnel.
2. Put five spoonfuls of soil into the funnel.
3. Place the funnel into the top of a jar.
4. Slowly pour water into the soil.

There are lots of different types of soil.

Chalky soil	Sandy soil
• Light brown with white pieces • Lots of holes, full of air • Water drains quickly	• A light colour • Lots of holes, full of air • Water drains quickly • Feels dry
Loam	**Clay**
• Dark brown • Ideal mixture of sand, clay and dead animals and plants • Water drains well • Full of the chemicals needed by plants	• Orange or grey • Very few holes, so not much air • Water drains slowly • Feels damp or wet

✏️ Which soil will get muddy after heavy rain?

✏️ Which soil will be too dry to grow crops?

✏️ Circle the correct answer.

1 Rocks are used for buildings because they are   soft   hard
2 Small, round pieces of rock are called   granite   pebbles
3 The best soil for growing crops is   loam   chalky

Now turn to page 69 to review what you know.

What we have learned about looking at rocks

What are rocks? (pages 56–7)

✎ Why does rock make a good building material?

I know that all rocks were formed millions of years ago. ○

I understand how the hardness of a rock makes it useful for some things but not others. ○

Different building materials (pages 58–9)

✎ Thousands of years ago in the Middle East many people's houses were made of clay. What made clay easier to use than stone?

✎ Some building materials like rocks, clay and wood are natural. Name three man-made building materials.

I can name three materials that are completely man-made. ○

I can name three materials that are completely natural. ○

Types of rock (pages 60–1)

✏️ Name two rocks that were made millions of years ago.

✏️ Which rock is made from seashells?

I can name five kinds of rock. ◯

How do we use rocks? (pages 62–3)

✏️ Which rock would you use to build the steps of a building?

I understand why different rocks are used for different purposes. ◯

Different types of soil (pages 64–7)

✏️ Soil with a lot of chalk in it is known as chalky soil. Name two other kinds of soil.

I know that only some soils are good for growing plants. ◯

5 Day and Night

In this module you will:

- explore how the Sun appears to move during the day
- find out how shadows change
- discover how the spin of the Earth leads to day and night.

Look at the photograph of the Sun. What words would you use to describe it?

Did you know that the Sun is actually a star? It is the biggest star in the sky. The Sun is bigger than 1 million Earths. Can you imagine that?

The Sun is our main source of light.

Word Cloud

compare
Earth
night
Sun
measure

Physics

- What happens to the Sun at night?
- Look at the photograph of the night sky. How dark is dark? What do we mean when we say it is dark?

Amazing fact

The hottest temperature recorded on the surface of the Earth was 60°C. The temperature on the surface of the Sun is 6000°C. That's hot!

Day and Night

71

The Sun appears to move in the sky

Explore whether the Sun moves during the day. Discover how shadows change.

The Big Idea

The Sun does not move.

💬 What does the Sun look like in the morning?

In the morning the **Sun** gives us light. It is the main source of light on **Earth**. During the day the Sun appears to get higher in the sky. From midday the Sun appears to go lower and lower until it disappears.

💬 What happens when the Sun goes down?

✏️ Where do you get light from at night?

People say that the Sun rises in the morning and goes down at **night**. This is confusing because it makes us think that the Sun moves across the sky. This is not true.

It is the movement of the Earth that gives us day and night, not the movement of the Sun.

Look at the picture of the Earth. The axis of the Earth is an imaginary line through the Earth from the North Pole to the South Pole. The Earth spins on its axis. This takes 24 hours.

✏️ How many hours are there in a day?

For half of the time it takes the Earth to spin, half of the Earth is in sunlight.

💬 What happens for the other half of the time that the Earth spins?

This is how we get day and night on Earth.

The Sun appears to move in the sky

Explore whether the Sun moves during the day. Discover how shadows change.

The Big Idea

The Sun does not move.

Investigation: The Sun survey

> ⚠ Do not look directly at the Sun – it might hurt your eyes. In summer and at midday the Sun is very powerful. It can burn our skin even though it is far away.

1 Go outside in the morning and look at where the Sun is.
2 Is the Sun very bright? Is it hot? Can you feel the Sun on your skin?
3 Go back to the same place at different times of the day.

💬 Did the Sun appear to move across the sky?

✏ When was the Sun brightest?

✏ At what time was the Sun above you?

In the morning and evening the Sun is at its lowest points in the sky. At midday the Sun is at its highest point.

✏️ Look at the picture. When is the Sun at its highest?

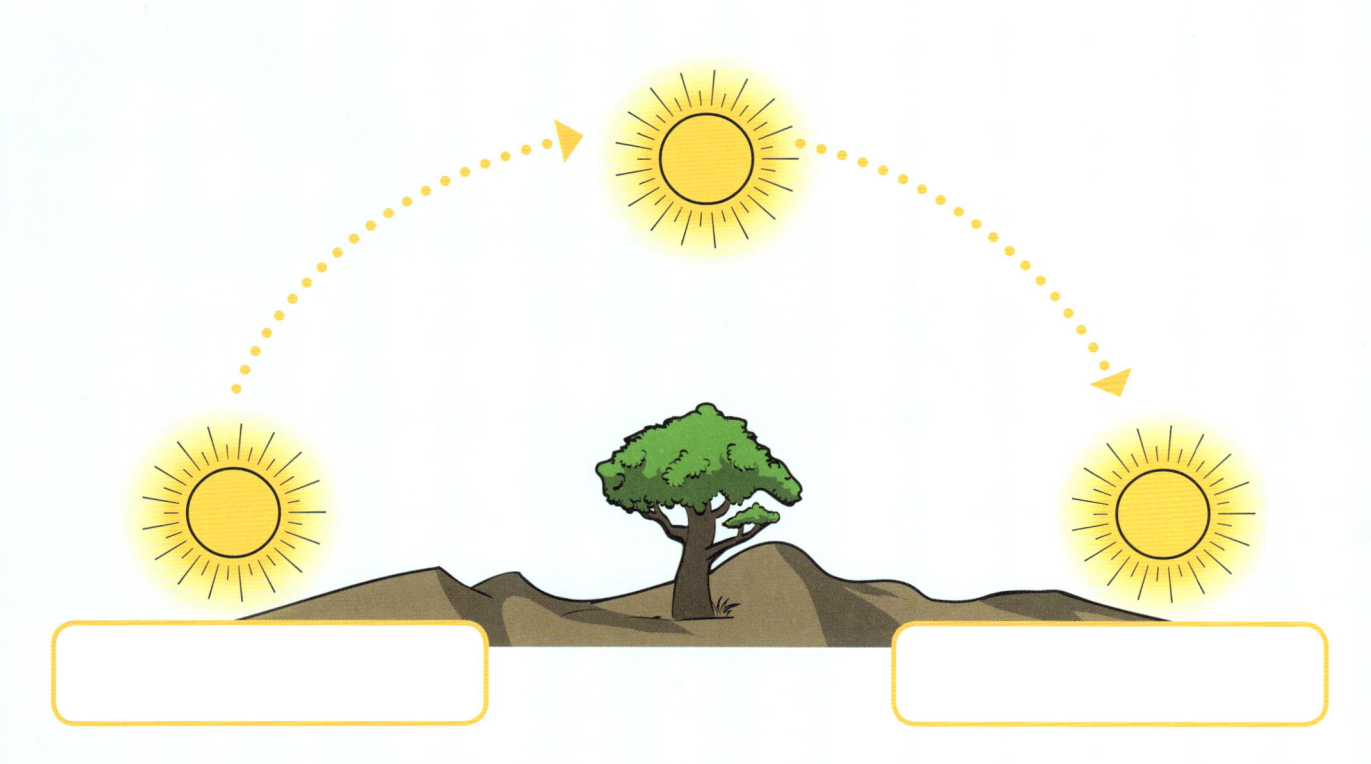

✏️ Use the word bank to complete the labels on the picture. When is it evening? When is it morning?

Word Bank

Evening   Morning

Think about...
When is the Sun hottest? When are we most at risk from sunburn?

Day and Night

75

Now turn to page 80 to review what you know.

Shadows change during the day

Explore whether the Sun moves during the day. Discover how shadows change.

The Big Idea

Shadows move.

💬 Can you remember how a shadow is formed?

A shadow is formed when an object blocks the light. A shadow is the dark black shape of the object that is blocking the light.

Remember
The Sun is not moving. The Earth is spinning on its axis.

✋ **Investigation: Exploring hand shadows**

1 Hold a torch very close to your partner's hand.
2 Ask your partner to form a shadow with their hands.

💬 Look carefully at the shadow. How long is it?

3 Now move the torch further away from your partner's hand.

💬 What happens to the length of the shadow?

🗨 When you did your survey of the Sun, did you notice your own shadow?

✋ **Investigation: Exploring your own shadow**

1 Choose somewhere to stand. Stand with your back to the Sun.

2 Ask your partner to measure the length of your shadow.

🗨 Look at your shadow. Is it long and thin or short and fat?

3 Measure your shadow again at midday, and later in the afternoon.

🗨 Is there a pattern in your results?

✏️ How does your investigation prove that the Sun is at its highest at midday?

✏️ Circle the correct answer.

1	The Sun moves across the sky.	True	False
2	At midday the Sun is low in the sky.	True	False
3	When the Sun is high in the sky a shadow is short.	True	False
4	Shadows change during the day.	True	False

Now turn to pages 80–1 to review what you know.

The Earth is spinning

Show how the spin of the Earth leads to day and night.

The Big Idea

Help, the Earth is spinning!

The Sun lights up part of the Earth when the Earth spins

💬 Why do some people think that the Sun moves across the sky?

✏️ How do we get day and night? Use the word bank to complete the sentences.

The Earth _____ so one part is facing the _____.
The part is in _____. The other side of the
Earth is away from the Sun. This part is in _____.

Word Bank
Sun   daytime   night-time   spins

Remember
The Earth is spinning on its axis.

✏️ How many hours does it take the Earth to make one complete spin?

✋ **Investigation: Spinning Earth**

We can pretend to be the Earth and the Sun with a torch and a ball. The ball can be the Earth.

💬 What could the torch be?

1. Work with a partner. Choose to be either the Sun or the Earth.
2. Put a mark on the ball to show where we are on Earth.
3. Hold the torch very, very still!
4. Hold the Earth in front of the Sun and slowly turn it.

Think about...
It takes 365 and a quarter days for the Earth to move around the Sun. What else do you know that has 365 days?

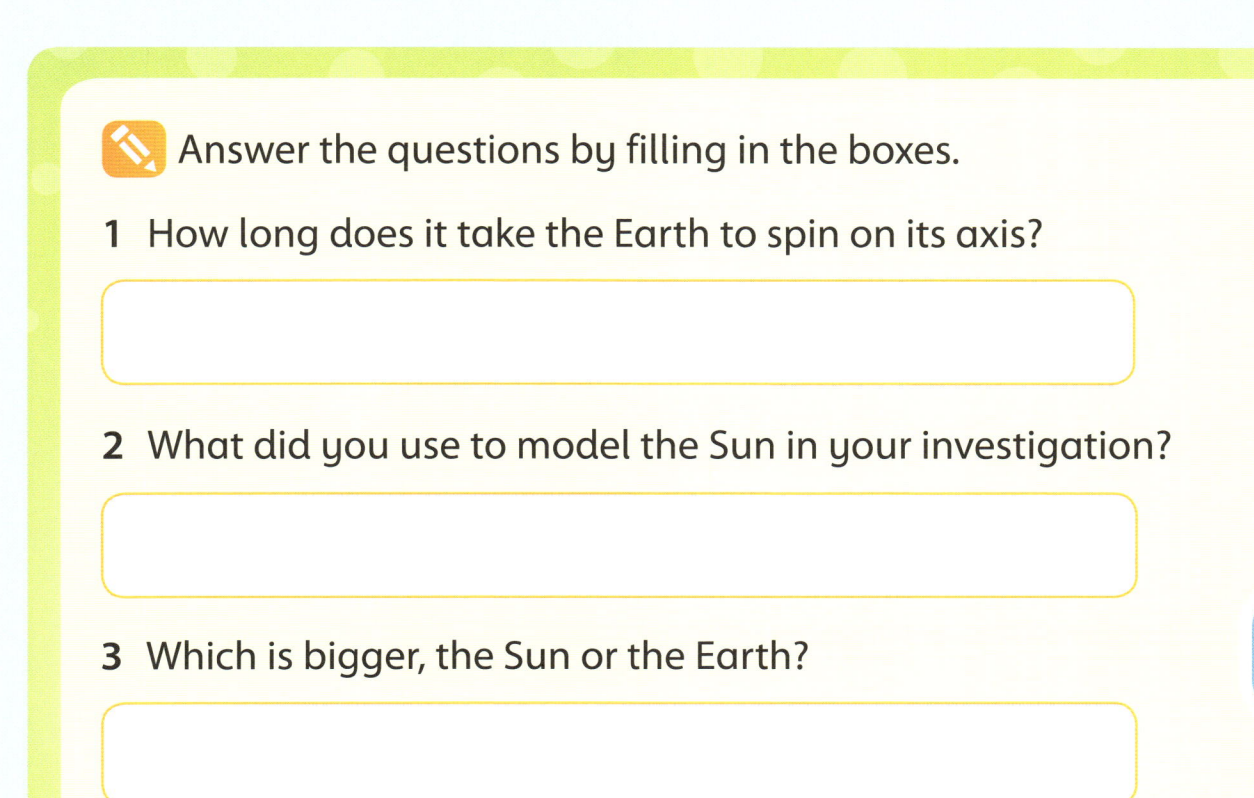

✏️ Answer the questions by filling in the boxes.

1. How long does it take the Earth to spin on its axis?

2. What did you use to model the Sun in your investigation?

3. Which is bigger, the Sun or the Earth?

Now turn to page 81 to review what you know.

What we have learned about day and night

The Sun appears to move in the sky
(pages 72–5)

✎ At what time of day does the Sun appear highest in the sky?

✎ At what time of day does the Sun appear lowest in the sky?

I know that the Earth turns and the Sun does not move across the sky. ○

I understand why some people think that the Sun moves across the sky. ○

I understand why the Sun seems lower in the sky in the morning and at night. ○

I can describe why the Sun is hotter at midday. ○

Shadows change during the day (pages 76–7)

✎ What is a shadow?

✏️ At what time of day are shadows longest?

✏️ Why are shadows shortest at midday?

I know that shadows are formed when light is blocked. ○

I understand why shadows change length during the day. ○

The Earth is spinning (pages 78–9)

✏️ What happens to the sky when your country is facing away from the Sun?

✏️ Why does your country not face the Sun all the time?

I understand how the spin of the Earth creates day and night. ○

6 Plants and Animals

In this module you will:

- identify some of the plants and animals that live in hot, dry climates
- learn how plants and animals keep cool and stay alive
- look at the different environments and their plant and animal life
- look at the different types of weather
- learn how hot, dry countries get their water.

Can you see from the picture that parts of the Earth are almost yellow? This is because there is hardly any water.

Amazing fact

Part of the Western Desert of Egypt did not see a drop of rain for 17 years!

Biology

There are millions of different kinds of plants and animals in the world. In the Middle East there are only a few thousand. The most common tree growing wild is the winter thorn tree.

Q The Asiatic cheetah is the fastest land animal. It can reach speeds of 120 kilometres per hour! There are only a few left in the Middle East. Why do you think this is?

Q What wild plants and animals do you find where you live?

animal
hot   plant
rainy   sunny
weather   windy
wet

Word Cloud

Plants and Animals

83

Plants and animals in hot countries

Understand that there aren't many different types of plants and animals in the Middle East.

The Big Idea

It is difficult to live and grow in places that are hot and dry.

Do you remember the list you made of the **plants** and **animals** where you live? Look at the pictures. These are some plants and animals that you may have seen.

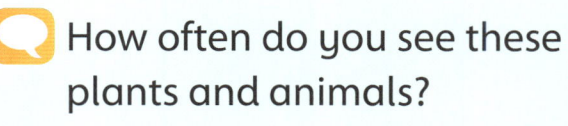

- Camel
- Goat
- Lizard
- Snake
- Locust
- Gerbil
- Moth
- Butterfly
- Finch
- Date palm
- Olive tree
- Punarnva flower
- Desert rose
- French tamarisk

💬 How often do you see these plants and animals?

84

In countries like England you can see 30 to 50 different kinds of plants and animals on any day.

Q Look at the pictures. Where would you choose to live?

Look at the words below. Circle three words that make living in the city easy.

Underline three words that make living in the desert difficult.

- hot
- water
- shade
- no water
- cool
- no shade

Think about...
In dry and hot countries plants and animals also need water and to keep cool. How do you keep cool and save water?

Plants and Animals

Plants and animals in hot countries

Learn how plants keep cool and save water.

The Big Idea

Living things in very hot conditions die without water.

There are not many plants that can stay alive when the **weather** is very **hot** and dry. Those that can have clever ways of keeping cool and saving water. Unlike animals, plants cannot move to find shade and water!

If plants are lucky, their seeds fall in shady places.

Some plants, like juniper and eucalyptus, have adapted to grow in full sunlight.

Shady oasis

Juniper

Eucalyptus

Some plants, like palm trees, develop leaves with thick tough skins that protect them from the Sun. Some leaves also stop water escaping from the plant.

✋ **Investigation: How plants stop water being lost from their leaves**

1. Wet three pieces of cloth.
2. Copy the pictures.
3. Put them outside in the Sun until the open cloth is dry.

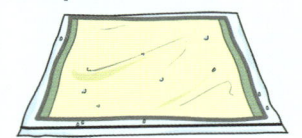

Wet cloth in sealed plastic bag

Open wet cloth

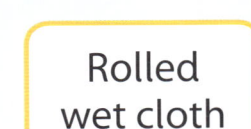

Rolled wet cloth

💬 What did you find?

✋ **Investigation: What happens when a plant has no water?**

1. Put the potted plants outside in the Sun.
2. Do not water plant number 2.
3. Observe the plants over the next few days.

💬 What differences did you see?

Plants and animals in hot countries

Learn how animals keep cool and save water.

The Big Idea

Living things in very hot conditions die without water.

There are not many **animals** that can stay alive when the weather is very hot and dry. Those that can have some very clever ways of keeping cool and saving water.

Animals like elephants and camels have thick skins to protect them from the Sun.

Animals like lizards, gerbils and snakes hide from the Sun in burrows or under rocks.

Animals like bats and moths only come out at night.

✏️ Do you remember your work about the Sun? Circle the time of day when you would prefer to be outside? Why is this?

| sunrise | midday | sunset |

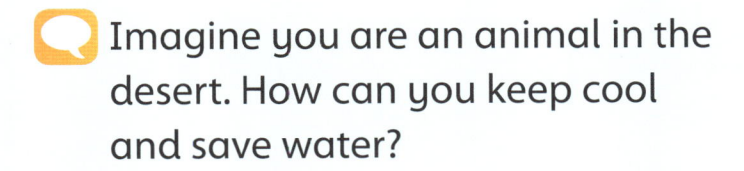

💬 Imagine you are an animal in the desert. How can you keep cool and save water?

Think about...
Did you know that a camel can save litres of water mixed with fat in its hump? This water lasts for days!

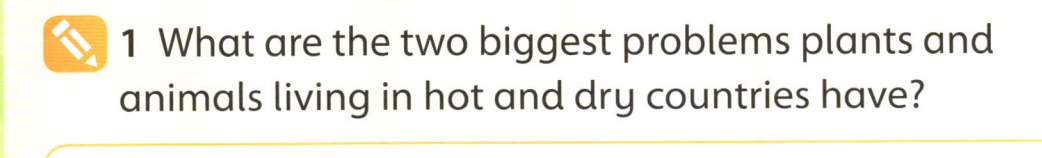

✏️ **1** What are the two biggest problems plants and animals living in hot and dry countries have?

2 What can plants do to stay strong and healthy?

3 What can animals do to stay strong and healthy?

Now turn to page 100 to review what you know.

Plants, animals and their environment

Understand that different plants and animals live in different environments.

The Big Idea

How the environment looks depends on how much water there is.

The environment is the land and climate where plants and animals live. In hot, dry countries there is less and less water the further away you are from the sea.

Remember
The more water there is, the more plants and animals there are.

✏️ Look at the pictures. Use the word bank to show what animals and plants could live in these environments. An example for each has been done for you. If you know of other animals you might see, you can add those too.

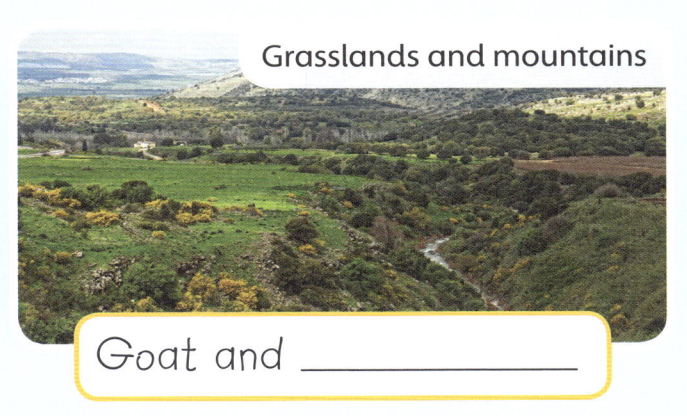

Warm sea

Tuna and _____

Coastal fog woodlands

Warbler and _____

Wetlands

Reeds and _____

Grasslands and mountains

Goat and _____

Desert

Camel and _____

Word Bank

turtle   oryx   scorpion   sheep   toad

Plants and Animals

Plants, animals and their environment

Understand about caring for the environment.

The Big Idea

Not caring for our environment kills plants and animals and pollutes the water.

Riyadh skyline at night

We share our environment with plants and animals, so we need to take care of it.

Did you know that when there are more people in the world, pollution gets worse? The city of Riyadh has grown from 150 000 inhabitants in the 1960s to over 5 million today.

✏️ There are lots of different types of pollution. Circle three.

> litter   traffic
> noise   rubbish   smog
> river   mountain   town
> city

Wadi Hanifa

💬 Look at the picture. This area used to be so full of rubbish no one could enjoy it. How can we stop it being spoiled again?

✋ **Investigation: Where is the pollution?**

1 Look around your school. Can you see any litter?
2 What can you hear? Is there any noise pollution?
3 Look around on your way home from school. What do you notice?

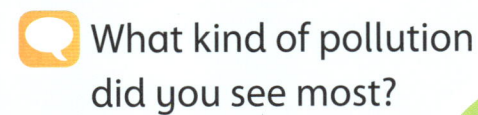

💬 What kind of pollution did you see most?

Amazing fact

Did you know that in the Asir Mountains in the Arabian Peninsula there is a colony of baboons in the city? They followed the tourists who came and fed them. They find it easier to live in a city where there is food and water.

✏️ 1 Name three different environments that can be found in hot and dry countries.

2 Name one thing that you can do so you don't pollute your environment.

Now turn to pages 100–1 to review what you know.

Weather and water in hot countries

Understand the changes in the weather.

The Big Idea

In hot, dry countries the weather only changes from season to season.

A dust storm crossing the Red Sea

In hot, dry countries the weather only changes when the seasons change. In the summer it can be a very hot 40–50°C in the day and 20°C at night.

In the winter it can be a warm 15–20°C in the day and between 5°C and freezing at night.

The only time it rains, and that's not very often, is in the winter.

✏️ The weather is often described in opposites. Draw a line between the words in the bubble that mean the opposite of each other. One has been done for you.

💬 Can you think of any symbols that you could draw for '**sunny**', '**cloudy**', '**windy**' and '**rainy**'?

> wet calm
> dry hot
> cold clear stormy
> cloudy

✋ **Investigation: My weather this week**

1 Measure and record the weather for one week each season.

2 Use weather symbols to show what you have seen.

Think about...
Why do you think the nights are cool in hot, dry countries?

Amazing fact
The Khamsin is a hot, dry wind in the Western Desert of Egypt. It can reach speeds of up to 150 kilometres per hour. That is faster than the top speed of most cars!

Plants and Animals

Weather and water in hot countries

Understand where the water comes from in hot, dry countries.

The Big Idea

In hot, dry countries water is precious.

In some countries it rains a lot!

In hot and dry countries it rains less than 100 millimetres per year. In England it rains ten to twenty times that amount. That's quite a difference!

Water that comes from the sky is known as rainwater.

Water also comes from under the ground. The ground soaks up the rainwater. This is known as groundwater. This water can be reached by drilling wells.

Did you know that you can find water in the middle of a desert?

💬 Where does the water come from?

Water is taken from the oasis by the heat of the day and by people and animals, but it remains full of water year after year.

💬 Think about the water you use at home for washing up and the water you use at school. How do you think it gets there?

An oasis in the desert

✏️ Choose one thing you do at home that uses water. Where did the water come from?

✏️ Draw a picture to show where the water came from.

Weather and water in hot countries

Understand where the water comes from in hot, dry countries.

The Big Idea

We cannot drink the water that comes directly from the ground or the sea.

✋ **Investigation: Cleaning groundwater**

Sometimes groundwater can be dirty and muddy, so it needs to be cleaned.

1 Take some muddy water.
2 Stretch a piece of cloth over a bowl.
3 Pour the muddy water through the cloth.

✏️ What has happened to the water?

✏️ What has happened to the mud?

The water has been filtered. This is called filtration.

⚠️ Do not drink the water. All filtered water needs to be boiled in case there are any bugs in it.

Hot, dry countries also use sea water for the water they need. This water is very salty and is poisonous if you drink a lot of it. It also needs cleaning.

✋ **Investigation: Cleaning sea water**

1 Fill a glass dish with salty water.
2 Put it in the Sun and wait for the water to disappear.
3 Look carefully at the dish. What do you see?

💬 Where has the water gone?

✋ **Investigation: Capturing drinking water from salt water**

1 Put some salt water in a bowl. Mark how high the water is.
2 Cover it with a piece of glass and put it in the Sun.

💬 Has the water disappeared?

✏️ 1 In what season does it rain in hot, dry countries?

2 Name two sources of water that we use for drinking.

3 Why can't we drink sea water straight from the sea?

Now turn to page 101 to review what you know.

What we have learned about plants and animals

Plants and animals in hot countries (pages 84–9)

✏️ Name two ways that plants keep cool.

✏️ Name two ways that animals keep cool.

I understand how plants and animals keep cool in very hot countries. ◯

Plants, animals and their environment (pages 90–3)

✏️ Name three different environments that you can find in hot, dry countries.

✏️ Name two animals that live in the desert.

✏️ In a growing country, what is the biggest threat to the environment?

I know that different environments have different kinds of plants and animals. ◯

I know that we need to care for our environment. ◯

Weather and water in hot countries (pages 94–9)

✎ How often does the weather change in hot, dry countries?

✎ Name the three sources of water to be found in hot, dry countries.

I know that in hot, dry countries the weather only changes from season to season. ◯

I know three sources of water to be found in hot, dry countries. ◯

Glossary

Key words

animal

battery

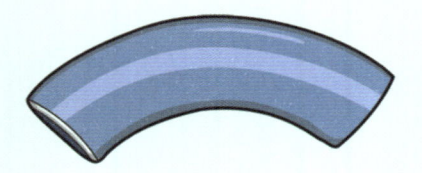

bend

102

bulb

circuit

compare

Glossary

103

dark

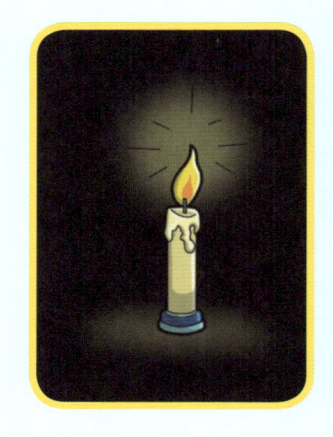

Earth

guess (predict)

104

hot

light

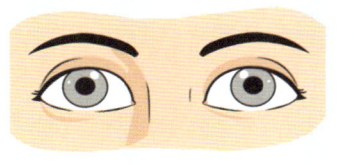

look

material

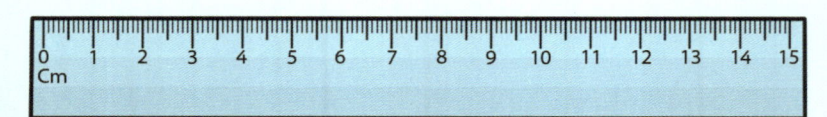

measure

melt

natural

night

plant

prove

rainy

Glossary

107

rock

sandy

soil

solid

stone

stretch

Sun

sunny

switch

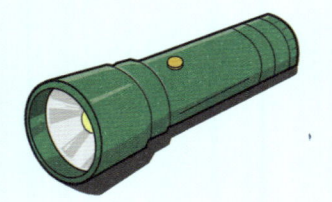

weather

torch

windy

twist

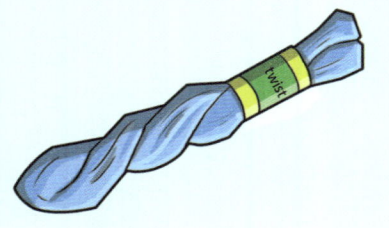

wire

110